AF802861

Hobbes and the Making of Modern Political Thought

Continuum Studies in British Philosophy
Series Editor: James Fieser, University of Tennessee at Martin, USA

Continuum Studies in British Philosophy is a major monograph series from Continuum. The series features first-class scholarly research monographs across the field of British philosophy. Each work makes a major contribution to the field of philosophical research.

Applying Wittgenstein—Rupert Read
Berkeley and Irish Philosophy—David Berman
Berkeley's Philosophy of Spirit—Talia Bettcher
Bertrand Russell, Language and Linguistic Theory—Keith Green
Bertrand Russell's Ethics—Michael K. Potter
Boyle on Fire—William Eaton
The Coherence of Hobbes's Leviathan—Eric Brandon
Doing Austin Justice—Wilfrid Rumble
The Early Wittgenstein on Religion—J. Mark Lazenby
F. P. Ramsey—edited by Maria J. Frapolli
Francis Bacon and the Limits of Scientific Knowledge—Dennis Desroches
Hume on God—Timothy S. Yoder
Hume's Social Philosophy—Christopher Finlay
Hume's Theory of Causation—Angela Coventry
Idealist Political Philosophy—Colin Tyler
Iris Murdoch's Ethics—Megan Laverty
John Stuart Mill's Political Philosophy—John Fitzpatrick
Matthew Tindal, Freethinker—Stephen Lalor
The Philosophy of Herbert Spencer—Michael Taylor
Popper, Objectivity and the Growth of Knowledge—John H. Sceski
Rethinking Mill's Ethics—Colin Heydt
Russell and Wittgenstein on the Nature of Judgement—Rosalind Carey
Russell's Theory of Perception—Sajahan Miah
Thomas Hobbes and the Politics of Natural Philosophy—Stephen J. Finn
Thomas Reid's Ethics—William C. Davis
Wittgenstein and Gadamer—Chris Lawn
Wittgenstein and the Theory of Perception—Justin Good
Wittgenstein at his Word—Duncan Richter
Wittgenstein on Ethical Inquiry—Jeremy Wisnewski
Wittgenstein's Religious Point of View—Tim Labron

Hobbes and the Making of Modern Political Thought

Gordon Hull

continuum

Continuum International Publishing Group
The Tower Building
11 York Road
London, SE1 7NX

80 Maiden Lane
Suite 704
New York, NY 10038

www.continuumbooks.com

© Gordon Hull 2009

All rights reserved. No part of this publication may be reproduced or transmitted in any form or by any means, electronic or mechanical, including photocopying, recording, or any information storage or retrieval system, without prior permission in writing from the publishers.

British Library Cataloguing-in-Publication Data
A catalogue record for this book is available from the British Library.

ISBN: HB: 978-1-4411-4002-9

Library of Congress Cataloging-in-Publication Data
Hull, Gordon, 1972-
Hobbes and the making of modern political thought/Gordon Hull.
p. cm.
Includes bibliographical references and index.
ISBN-13: 978-1-4411-4002-9 (HB)
ISBN-10: 1-4411-4002-6 (HB)
1. Hobbes, Thomas, 1588-1679. 2. Biopolitics. 3. Political science–Philosophy.
I. Title.
B1247.H85 2009
320.1092–dc22 2009007849

Typeset by Newgen Imaging Systems Pvt Ltd, Chennai, India
Printed in Great Britain by the MPG Books Group, Bodmin and King's Lynn

Contents

Preface

This book is a genealogical narrative about the emergence of "modern" political philosophy. Both the rough outlines of the story—during the seventeenth century, "social contract theory" emerges, replacing "Aristotelianism"—and the names of many of the main players will be familiar. In the seventeenth century, in the history told by the victorious moderns, this story had all the trappings of a melodrama: Aristotle and his followers represented a monolithic horde of bad philosophical reasoning, the adoption of which had catastrophic effects on mind and city; the heroic restart of reason along proper lines held out the hope of rescuing humanity from its intellectual darkness. At the present historical juncture, as we become increasingly unsure of our own modernity and even of what the claim to be "modern" entails, while at the same time supposedly transitioning to a late- or postmodern condition, these seventeenth-century narratives deserve reconsideration.

A great deal of work has already been done in this regard, on Descartes in particular. My hope is that the present work will contribute to Hobbes's inclusion in our reconsideration of the seventeenth century, and in understanding how philosophical positions carved out in that century continue to demand our attention today. Several intertwined threads are recurrent in the study: the elevation of construction (*poiesis*) as a metaphor for thought at the expense of classical models of intellection; the emergence of quantification (often in the guise of geometry), and in particular a growing willingness to treat subjectivity as quantifiable; the breakdown of the Aristotelian nature/art distinction, at the expense of nature; the decline of *phronesis* as a form of practical reason; the reconfiguration of the Aristotelian *anima*, which in Hobbes's case involves a materialist psychology that elevates the importance of imagination; and the general ascendance of nominalist theses about language and mind. I begin with our ongoing fascination with Hobbes, and conclude by proposing what I take to be a causal reason for this fascination: not only is our age biopolitical but also Hobbes himself is among the first thoroughly biopolitical thinkers.

Writing on Hobbes induces a certain humility, not just because he is such a formidable thinker but also because so many others have written before. Many of these debts are recorded in the notes and bibliography, and without them, this book would not have been possible. My personal debts, incurred in conversations over the years, are sufficiently numerous that any attempt to record them all would fail. I would, however, like to explicitly acknowledge the generosity of a few individuals whose support of the project, and of my work more generally over the years, has gone well beyond the call of duty: Jay Bernstein, Idit Dobbs-Weinstein, Gregg Horowitz, Warren Montag, Tim Reiss, and Tony Smith. My debt to Maya Socolovsky, my constant interlocutor and partner, is beyond what can be measured. Finally, my children, Ilan and Amia, both help me keep academic work in proper perspective and serve as a reminder of why it matters what kinds of worlds are dreamt of in our philosophies.

An earlier version of Chapter 3 appeared as "Hobbes and the Pre-Modern Geometry of Modern Political Thought," in *Arts of Calculation: Numerical Thought in Early Modern Europe*, ed. David Glimp and Michelle Warren (St. Martins/Palgrave, 2004), pp. 115–35. An earlier version of Chapter 4 appeared as "Hobbes's Radical Nominalism," *Epoché* 11 (Fall 2006), pp. 201–23.

Abbreviations and a Note on Texts

Where possible, I follow convention in referring to historical sources (thus, for example, Bekker page numbers for Aristotle, chapter and paragraph for most of Hobbes's works, etc.). Departures from published translations are my own, and may be referenced to the original text. Translations of otherwise untranslated sources are my own. I also use the following abbreviations, for which full citations are provided in the bibliography:

Hobbes

DC: *De Cive*, by chapter and paragraph; I generally follow the translation by Tuck and Silverstone, *On the Citizen*

DCo: *De Corpore*, by chapter and paragraph

DH: *De Homine*, by chapter and paragraph; I generally follow the translation in Gert, *Man and Citizen*

DM: *De Motu*, by chapter and paragraph

EL: *Elements of Law, Natural and Politic*, by chapter and paragraph

EW: *The English Works of Thomas Hobbes*, by volume and page

HC: *Correspondence of Thomas Hobbes*, by page

L: *Leviathan*. By chapter and paragraph (in the Curley edition), followed by page (from the Tuck edition). I retain the original spelling from Tuck

OL: *Opera Philosophica quae latine scripsit omnia*, by volume and page

Other Authors

2T: Locke, *Second Treatise on Government*, in *Two Treatises on Government*, by paragraph

AT: Descartes, *Oeuvres de Descartes*, by volume and page

NE: Aristotle, *Nichomachean Ethics*, by Bekker pagination

Op Th: William of Ockham, *Opera Theologica*, by volume and page

P: Machiavelli, *The Prince*, by chapter and page
PWD: Descartes, *Philosophical Writings of Descartes*, by volume and page
SL: William of Ockham, *Summa Logica.* I generally follow the translation in Loux, *Ockham's Theory of Terms*
STh: Thomas Aquinas, *Summa Theologiae*
W: Machiavelli, *The Chief Works and Others*

Chapter 1

Introduction: The Politics of Construction

And how Capital a Delinquent is Mr. Hobs, who hath by writeing endeavoured to render the sentiments of the best and most learned ridiculous?

John Dowel

1.1 Hobbes, Our Contemporary

Hobbes, it seems, is everywhere. His description of life in the state of nature as "nasty, poore, solitary, brutish, and short" rivals Descartes's *cogito* for recognition outside philosophy departments, and for inclusion in introductory syllabi within them. There are numerous editions of the 1651 English *Leviathan* in print, and recent or forthcoming editions and translations of many of his other works. The secondary literature is vast and growing; Edwin Curley (1989) remarked 20 years ago that the scholarship had reached a level of sophistication such that Hobbes specialists were describing a thinker almost unrecognizable in the caricatures that are presented as part of one's standard philosophical training; that observation is certainly even more true today. At the same time, explanation of *why* study of Hobbes should be so compelling seems to have reached somewhat of an impasse. This question is worth at least an initial puzzlement: Hobbes advocates a form of government that almost no one in contemporary Western democracies would endorse, and denunciations of Hobbes or Hobbesianism have been a cottage industry since shortly after his writings appeared. As early as 1673, one Dr. John Templar was moved to declare that Hobbes was the "Malmesburian Hydra, the enormous Leviathan, the gigantic dragon, the hideous monstrosity and British beast, the Propagator of execrable doctrines . . . the Nonsensical roguish vendor of falsifications."[1] Reading Hobbes did not enlighten—"he seems to stray from the truth in every single claim which he advances as his own," Descartes opined (HC 57)—and could actually damage; John Dowell concluded in 1683 that "No person that hath suckt in *Hobbs* his Principles, can be a loyal subject" (p. 142). Variations of this view continue to the present; writing at the midpoint between Hobbes's time and our own, Marx tersely observed that, in Hobbes, "materialism becomes misanthropic" (1927, p. 305).[2] We are less inclined to think that reading a book can be damaging than were seventeenth-century thinkers, but it is still hard to find someone who says that Hobbes describes the *polis* as it ought to be.

So why read Hobbes at all? C. B. Macpherson had attempted to account for this state of affairs by situating Hobbes into the context of the emergence of capitalism. That thesis faced a violent end, and with it, apparently, perished much of the effort to understand Hobbes's recurrent fascination for us.[3] Charles Tarlton has recently suggested that Hobbes scholarship for the last 150 years or so has been involved in a careful sanitization of the "rigorous, blatant and elaborate apology for tyranny and arbitrary despotism to be found in *Leviathan*," and the slow transformation of Hobbes into the theorist of the modern state form (2001, p. 587). Tarlton's account seems importantly right—as I will argue, it is indeed Hobbes's modernity that fascinates us—but also leaves open the obvious question: why go to the trouble to make Hobbes one of us?

If Hobbesian absolutism has lost its luster, one aspect of Hobbes which does seem to resonate widely, in the form that Hobbes articulates it, without sanitization, is his description of the state of nature. I will cite two examples of this resonance.

(1) In a 1994 piece in *The Atlantic*, Robert Kaplan responded to the profusion of self-congratulatory, euphoric post-Cold War discourse by painting a darkly dystopic vision of a world plagued by population and environmental crises. Citing Thomas Fraser Homer-Dixon's work on the security implications of environmental problems, Kaplan moves seamlessly from Homer-Dixon to Hobbes. I quote him at length:

> [Homer-Dixon says,] "Think of a stretch limo in the potholed streets of New York City, where homeless beggars live. Inside the limo are the air-conditioned post-industrial regions of North America, Europe, the emerging Pacific Rim, and a few other isolated places, with their trade summitry and computer-information highways. Outside is the rest of mankind, going in a completely different direction."
>
> We are entering a bifurcated world. Part of the globe is inhabited by Hegel's and Fukuyama's Last Man, healthy, well fed, and pampered by technology. The other, larger, part is inhabited by Hobbes's First Man, condemned to a life that is "poor, nasty, brutish, and short." Although

> both parts will be threatened by environmental stress, the Last Man will be able to master it; the First Man will not. (1994, pp. 60–1)

The paper was quite influential: one commentator relates that "President Clinton was reported to have scribbled marginal notes on his personal copy, and citation of it became practically *de rigueur* for Cabinet members appearing before Congress."[4] That Kaplan's paper stuck a cultural chord should not distract us from its status as representative and therefore synechdochal. As Jennifer Widner notes, "to those who study Africa, Kaplan's essay . . . elicited a decidedly mixed reaction. The police-blotter description of familiar countries said nothing factually wrong, but it did not capture reality, either" (1995, p. 130). The problem was that the imagery obscured the details of conflicts and their origins, making them appear as simply intractable. In this regard, Kaplan's essay seemed a strategy for managing complexity which both drew attention to a phenomenon and at the same time hindered study of its details. I will return to strategies for managing complexity in the conclusion—and will contend that Hobbes's state of nature, like Kaplan's invocation of it, presents precisely such a discursive strategy. For now, I want to underscore the ready cultural availability of Hobbesian terms, and their resonance in describing the absence of social order and stability.

(2) Hobbes is the clear opposition point against which contemporary political theorists, such as Michael Hardt and Antonio Negri, often using the vehicle of Spinoza, attempt to theorize democracy. This is perhaps most evident in their *Multitude*, where they write:

> We are faced with a global state of war in which violence can erupt anywhere at any time. . . . The theory and practices of modern sovereignty were born by confronting this same problem, the problem of civil war—and here we are thrown back primarily to the seventeenth rather than the eighteenth century. Hobbes's reflections on the civil wars in England and Descartes's meditations on the Thirty Years' War in Germany are the founding moments of the dominant stream of modern European thought. Civil war is the negative instance against which the modern notion of political order is buttressed. The violent state of nature—the war of all against all—is really just a distilled, philosophical conception of civil war. (2004, pp. 238–9)[5]

The problem with Hobbes's solution, they suggest, is that it only redirects the war; "modern sovereignty . . . does not put an end to violence and fear but rather puts an end to civil war by organizing violence and fear into a coherent and stable political order" (2004, p. 239). On this reading, Hobbes is precisely the theorist of the modern state form, a label invoked not to sanitize Hobbes, but to show the deeply violent and antidemocratic character of the modern state. Working out of a very different problematic, Agamben similarly credits Hobbes with the founding insight of the modern political order; "the absolute capacity of the subjects' bodies to be killed forms the new political body of the West" (1998, p. 125).

These indicia from both philosophical and broader cultural discourse thus point in the same direction: Hobbes's problems are our problems; we not only face problems of civil war and instability but also find Hobbes attractive in our efforts to theorize those problems. In Hobbes we find a theorist haunted by the fragility of human life and political order, and in ourselves we find the worry that the Westphalian state system that first began to emerge in Hobbes's time is inadequate as a response to that fragility. The specter of Hobbes, then, registers a doubled anxiety about the *polis*. The simultaneous emergence of these anxieties is symptomatic of the cultural moment that ties us to Hobbes. After all, it is not the case that no one before or after Hobbes had thought about civil war; as I will argue (chapter 5.1), Hobbes's own account of the state of nature draws on and was effective partly because of its ability to resonate with a number of other readily available images, such as Thucydides' graphic description of the Corcyraean sedition. It is also not the case that no one before or after Hobbes has written about absolutism.

Why, then, Spinoza/Hobbes? Here, the concern is that later theorists take too much for granted, assuming without question too many tropes of philosophical modernity. A thinker like Kant, the argument goes, makes too many of the depoliticizing and antidemocratic moves of modern philosophy, especially now, when many of those moves are themselves being brought into question. If we want to pose the problems of democracy, we need to understand them through a period when the basic conceptual space of modern philosophy was being debated, not taken for granted. If Spinoza is the hero of this narrative, Hobbes is the enemy revenant. In Hobbes, we find both the paradigmatically modern political philosopher, and a crucial, liminal figure. His philosophical distillation of the state of nature out of the various depictions of civil war available to him resonates with us because we are willing to make with Hobbes many of his fundamental philosophical moves. We share his sense that the successful construction of the *polis* will require successfully channeling human passions and affects; and we share his fears both that this project might fail and that failure would be catastrophic. Finally, like Hobbes, we are not sure that we believe the legitimating narratives of modern thought. Or, rather, like Hobbes, we do not think that the legitimating narratives of modern thought can be taken for granted.

My suggestion, then, is that part of the fascination of Hobbes lies in the combination of his being the quintessentially "modern" political thinker and our own historical position in "late modernity." In this regard, it is not just the state of nature that we have inherited, but also the social contract as its remedy. Even a summary glance at some of the most important figures in modern political thought suggests that one of its defining moves argues that the political order is founded by an originary or constitutive contract of all of its members with each other. Before the institution of this contract, people live in a "state of nature"; thereafter, they live in "civil society." The sheer banality of sentences that begin, "when man emerged from the state of nature . . ." is *prima facie* evidence that we need to read someone for whom the state of nature/social contract pairing was anything but banal.

Let's rehearse the basic argument. For Hobbes, the state of nature is *unspeakably* bad, and avoidance of the collapse of civil society into the state of nature is the endgame of politics, so important that it justifies the establishment of an all-powerful state apparatus:

> In such condition, there is no place for Industry; because the fruit thereof is uncertain: and consequently no Culture of the Earth; no Navigation, nor use of the commodities that may be imported by Sea; no commodious Building; no Instruments of moving, and removing such things as require much force; no Knowledge of the face of the Earth; no account of Time; no Arts; no Letters; no Society; and which is worst of all, continual feare, and the danger of violent death; And the life of man solitary, poore, nasty, brutish, and short. (L 13.9, 89)

Life in the state of nature reduces humans to worse than animals. To remedy this situation, to live as humans, it is necessary "to erect . . . a Common Power" such that individuals in the state of nature "conferre all their power and strength upon one Man, or Assembly of men, that may reduce all their Wills . . . unto one Will: which is as much to say, to appoint one Man, or Assembly of men, to beare their Person" (L 17.13, 120). In other words, people in the state of nature are to renounce their material individuality in order to achieve the security of civil society. Their individuality will be reintroduced afterward, in the development of commerce, industry, knowledge, and other arts. However, that individuality will be regulated in principle by its presence in the commonwealth, and the presence of the commonwealth makes all the difference. The state of nature description in *Leviathan* should thus be read alongside Hobbes's earlier comparison of life inside and outside a commonwealth: "to sum up: outside the commonwealth is the empire of the passions, war, fear, poverty, nastiness, solitude, barbarity, ignorance, savagery; within the commonwealth is the empire of reason, peace, security, wealth, splendor, society, good taste, the sciences and good-will" (DC X.1).

It is very difficult now to read Hobbes with the shock that his early readers must have experienced. First, if the degree to which one finds a concept intuitive is a function of how often one hears it repeated, then surely among the most intuitive propositions of premodern political philosophy was that "man" was the "political animal," driven by nature into political community with other people. Hobbes emphatically asserts the contrary, and he did not stop there. He openly derided Aristotle, produced idiosyncratic readings of Scripture that seemed to deny existence to spiritual bodies, made heavy use of introspection in the manner of the *novator* Descartes, and relied on (a highly problematic understanding of) geometry and definition as universal guarantors of precision and certainty. On the basis of all this, he then declared that his own *De Cive* originated civil philosophy (EW I, xii). To add insult to injury, he did all of these things in elegant and felicitous prose and with a widely acknowledged rhetorical skill. We might find the presence of elegant prose in political theory surprising, but the rest—the incoherence of the "political animal," the rejection of Aristotle and other premodern thinkers, the effort to get Scripture out of political life, the importance of introspective thought experiments to philosophy, and the fascination with mathematics and logic as guarantors of method—is business as usual, albeit a business that an increasing body of literature is bringing into question.

In such a juncture lies an opportunity for thought, although only if one exercises great care. Hobbes's texts should not be reduced to some sort of vulgarized whipping post, as has too often happened to those of Descartes.[6] To be in a period of late modernity, such that the fundamental assumptions of modernity are no longer taken for granted, is only theoretically emancipating if we try to understand what those modern assumptions are and how they came to be. It is the tremendous achievement of the materialist strands of Spinoza scholarship to draw our attention to the emancipatory potential of early modern thought for the present historical juncture. At the same time, Hobbes often appears as a nearly invisible point of opposition, against which the reconstruction of Spinoza's understanding of democracy is juxtaposed. It is time to bring Hobbes's modern materialism into full visibility.

1.2 Hobbes's Modernity

Let us then start with the most banal observation: "Hobbes was the founder of modern political philosophy."[7] Suppose we were to take this cliché seriously. To do so, we would need to know something about the adjective "modern," and something about fatherhood. Claims of parturition were in fact a commonplace of the seventeenth century. Descartes is perhaps the most famous example; he makes originality both a methodological precept and the basis of his own narrative of himself. This precept often arrives innocently enough. In the third of his early *Regulae*, for instance, he suggests that "we ought to read the writings of the ancients, for it is of great advantage to be able to make use of the labors of so many men." A few lines later, however, he begins attaching provisos to the advice: "there is a considerable danger that if we study these works too closely traces of their errors will infect us and cling to us against our will and despite our precautions" (PWD I, 13; AT X, 366). Even if we remained uninfected, he adds, and "even though we should have read all the arguments of Plato and Aristotle, we shall never become philosophers if we are unable to make a sound judgment on matters which come up for discussion" (PWD I, 13; AT X, 367). Thus, by the *Discourse on the Method*, the narrator, while cautioning that his procedure is "not an example that everyone ought to follow," nevertheless proposes that "my plan has never gone beyond trying to reform my own thoughts and construct them upon a foundation which is all my own" (PWD I, 118; AT VI, 15). Indeed, he reports that, "as soon as I was old enough to emerge from the control of my teachers, I entirely abandoned the study of letters" (PWD I, 115; AT VI, 9); and that, as he sat before the famous wood stove in Germany, "among the first" thoughts to occur to him was that "there is not usually so much perfection in works composed of several parts and produced by various different craftsmen as in the works of one man" (PWD I, 116; AT VI, 11). In the following paragraph he draws a political analogy, suggesting that states founded by one lawgiver, such as the Sparta of Lycurgus, were preferable to those founded on the accretion of custom. That this image of self-creation was entirely a fabrication focuses our attention on the metaphorics underlying it, according to which Descartes' version of being modern is best imaged as an act of self-creation. As Paul Valéry put it, "it would seem that, in everything, he took his Self, of which he was so powerfully aware, as the point of origin of the axes of his thought" (qt. in Lachterman, 1989, p. 126).

The image of a radical break with antiquity was recurrent among the moderns. It was extreme enough that Leibniz, by his 1686 *Discourse on Metaphysics*, had called for somewhat of a retreat, suggesting that "in the views of Scholastic philosophers and theologians there is much more of value than people suppose, provided they are used correctly and in their proper place" (1.11). Typical of the charge that there was literally nothing of use in Scholasticism, Hobbes's one-time employer, Sir Francis Bacon, proposed that "the whole operation of the mind must be completely re-started, so that from the very beginning it is not left to itself, but is always subject to rule; and the thing accomplished as if by machinery" (1994, p. 38). Bacon's comment is instructive both for its repetition of the general trope about starting over and for its blurring of the distinction between art and nature. As I will argue in the next chapter, this distinction was under considerable pressure by the seventeenth century, and images having to do with machines were proliferating culturally. In this regard, it is perhaps not too much of an exaggeration to propose that the most concise way of putting the Hobbesian strategy is as an effort to treat the *polis* as an object of *techné*, rather than according to nature.

Early modern political imagery also favored images of innovation, often with reference to states. Descartes's preference for lawgivers was not unusual, and echoes of it are found in several places, as for example Hobbes's sustained attack on the common laws in his late *Dialogue* on the subject. The figure most clearly implicated in political references to innovation is probably Machiavelli. Reference to Machiavelli was politically dangerous, and Bacon's comment that "we are much beholden to Machiavelli and other writers of that class, who openly and unfeignedly declare or describe what men do, and not what they ought to do" (1857, Vol. V, p. 17) stands out for its openness. Nonetheless, some of the rhetoric about innovation clearly echoes the Florentine. The case of Hobbes is quite difficult, and I will return to Hobbes's relation to Machiavelli in the closing section of Chapter 2. In the meantime, consider, for example, Descartes' invocation of Lycurgus with Machiavelli's praise in *The Prince* of "Moses, Cyrus, Romulus, Theseus, and others of that stamp" who "have become rulers through their own ability and not through luck or favor" (P VI, 20). That these men are to be imitated as examples indicates the instability of the advice: how can one imitate someone who was fully *sui generis*?[8] Similarly, in the beginning of the *Discourses on Livy*, Machiavelli compares himself to such innovators and explorers as Columbus and Amerigo Vespucci and claims that he has discovered a "path as yet untrodden by anyone" (Pfc. 1).[9] The advice again appears unstable, and he again retreats to the thought that he is really calling people's attention "to the lack of a proper appreciation of history"; his remedy is accordingly in the form of a commentary on Livy. There is in fact nothing new about claiming to be new, as Machiavelli seems to have borrowed the trope from Lucretius, who announces that "I traverse pathless tracts of the Pierides never yet trodden by any foot" (I.926–7). The language also looks forward to Descartes, who announces at the outset of the fourth of his *Regulae* that "so blind is the curiosity with which mortals are possessed that they often direct their minds down untrodden paths, in the groundless hope that they will chance upon what they are seeking" (PWD I, 15; AT X, 371). Descartes draws the lesson—and this is one of the lines that can be drawn to separate Machiavelli and the early moderns—that innovation requires a prior method.

Hobbes's own language shows continuities with Machiavelli, although it is clearly patterned along the lines of Descartes and Bacon in seeking both innovation and a method distinct from imitation of example. Thus, on the one hand, he flatly declares that "Reputation of Power, is Power" (L 10.3, 62). On the other hand, about the study of the ancients—in which his own achievements were considerable—Hobbes was generally dismissive. Although his attitudes on the subject evolve, by *Leviathan* he is able to argue that "the *vain-glory* which consisteth in the feigning or supposing of abilities in our selves, which we know are not, is . . . nourished by the Histories, or Fictions of Gallant Persons" (L 6.41, 42–3; I return to these passages in chapter 2). *Leviathan* similarly contains a number of programmatic dismissals of scholasticism, of which the following can be taken as emblematic. Scholastic physics, he writes:

> Was rather a Dream than Science, and set forth in senselesse and insignificant Language; which cannot be avoided by those that will teach Philosophy, without having first attained great knowledge in Geometry: For Nature worketh by Motion; the Wayes, and Degrees whereof cannot be known, without the knowledge of the Proportions and Properties of Lines, and Figures. (L 46.11, 461)

He adds that "I beleeve that scarce any thing can be more absurdly said in naturall Philosophy, than that which now is called *Aristotle's Metaphysiques*; nor repugnant to Government than much of that hee hath said in his *Politiques*; nor more ignorantly, than a great part of his *Ethiques*" (L 46.11, 461–2).[10] As with Descartes, these pronouncements should primarily be studied as examples of rhetoric, and in subsequent chapters I will try to be more precise about Hobbes's relation to certain strains of late scholasticism. I will also return in detail to his embrace of geometry, since, like Descartes, Hobbes thinks that mastering geometry is necessary to mastering other subjects; unlike Descartes, he enjoyed no lasting success as a geometer.

Hobbes's dismissals of scholastic and ancient thought were paired with the same sort of innovatory language as found in his contemporaries. For example, in the opening pages of his physics, he dates the beginning of "natural philosophy universal, which is the knowledge of the nature of *motion*" to Galileo, and proposes that

> Natural philosophy is therefore young; but Civil Philosophy is yet much younger, as being no older (I say it provoked, and that my detractors may

> know how little they have wrought upon me) than my own book *De Cive.* But what? were there no philosophers natural nor civil among the ancient Greeks? There were men so called . . . but it follows not that there was *philosophy.* (EW I, ix)

Hobbes thus clearly and deliberately positions himself with the moderns in their quarrel with the ancients. One thing these various post-Machiavellian claims about the novelty of philosophy have in common is that they claim to be autopoietic: philosophical progeny are said to spring complete from the minds of their creators, and we find none of the patient genealogies and enumerations of predecessors that mark the philosophical writing of the scholastics.

Such an immaculate conception of philosophy as proposed by the self-declared moderns would be unnatural in two ways. At the discursive level, it is profoundly disingenuous. As numerous studies have shown, the supposed clean familial lines demarcating scholastic from early modern thought were almost always in the service of hiding a considerable miscegenation. In Hobbes's case, the charge of unnaturalness is also substantive, or at least that is what I shall claim: Hobbes's political philosophy is at its most profound level a rejection of nature. Again, this claim requires both documentation and qualification. I will leave the documentation to the following chapters. The qualification is that Hobbes is rejecting a particular understanding of "nature." He rejects both natural teleology and the effort to base political theory on it. These rejections, I will argue, are consequent upon his understanding of thought in general and form the core of his understanding of politics. For Hobbes, the operative terms of politics are derived, not from physics or from theology, but, as he puts it in an early text, from "nothing but human wit" (EL 20.1).

By the time Hobbes wrote *Leviathan,* the ground had been thoroughly prepared for such a move. On the one hand, a century of revived skepticism and stoicism, developed against a background of brutal international warfare, had made it increasingly difficult to maintain the thesis that man is by nature a political animal. On the other hand, Machiavelli's declaration that a prince need only appear devout made it impossible to secure the terms through which an imitation of God could be regulated. Behind this declaration was the analogous thesis of the nominalists that knowledge itself was principally grammatical in the sense that thinking did not directly reflect more than its own operation on intuitive givens. On a different register, but at the same time, developments in science—not just Copernicanism, but Galileo's mechanics, developments in optics, and Bacon's almost giddy enthusiasm for productive works—made it possible to imagine different models of knowledge. The governing metaphor for scientific thought became thoroughly quantitative, and the early moderns—not just Descartes, but Hobbes and Spinoza—evidenced an unstinting enthusiasm for geometry.[11]

1.3 Lineages

The present book is a study of the coalescence of these lineages in the political thought of Hobbes. Insofar as Hobbes becomes the inaugural thinker of a distinctively "modern" way of thinking about politics, it is a study of the emergence of "modern" political philosophy and the assumptions which come to undergird it. I am thus interested in what one might call the discursive practice of early modern political thought, and in analyzing the various other discursive practices whose presence enables that thought.[12] By focusing on Hobbes, I adopt a conventional starting point, occasioned by our own academic and cultural invocations of Hobbes, and enabled both by the subsequent tradition of thought that names Hobbes as its progenitor, and by the sense—not just on the part of Hobbes himself—that the appearance of his work was in fact a significant event in political philosophy.

Hobbes's texts, I will argue, reflect and embody the early modern philosophical *ethos* more generally, as it emerges in the seventeenth century. By drawing this connection, I hope also to contribute to discussions about the proper specification of precisely what that early modern *ethos* is. In this regard, it is worth underscoring at the outset that although he is remembered today mostly for his contributions to political thought, Hobbes was at least as heavily invested in the same metaphysical, epistemological, and scientific debates that motivated thinkers such as Descartes.[13] The Hobbesian contribution to political philosophy needs to be seen through this lens. More specifically, Hobbes's innovation is the paradoxical combination of two elements. First, he understands politics as a productive art, rather than as a science in the traditional, Aristotelian sense. Second, this understanding, when coupled with a radicalized nominalism, allows him nevertheless to treat politics as if it were a demonstrative science. The figure of this innovation is the paired state of nature/social contract. To understand Hobbes's political project thus requires understanding his reinterpretation of Aristotelian science, in particular his understanding of knowledge, of the ordering of the sciences, and of the roles of logic, language, and geometry.

The book proceeds as follows. Chapter 2 consists of a series of analyses of discursive moments in the early seventeenth century that indicate the instability of some established patterns of thought relevant to reading Hobbes. I then show how these instabilities are manifest, usually self-consciously, in Hobbes's own texts. First, I look at some representative responses to Hobbes's work to underscore the extent to which that work broke with the terms of political discourse in early modern England. These responses find Hobbesian claims not just mistaken but incoherent. In those accusations of incoherence, I argue, lies evidence of a significant epistemic break. In other words, the problem with Hobbes was not the conclusions he drew; it was the method he used to get to them. Second, I attempt to unpack my sense of what that method is, that is, to explain and initially document my claim that Hobbesian thought is "constructive." A useful contrast, I will argue, is between Hobbes's texts and the Aristotelian taxonomy of the intellectual virtues outlined in *Nicomachean Ethics* VI.

Third, I look at the theoretical boundary between art and nature, which dated to Aristotle, and which was becoming increasingly difficult to sustain. Natural objects, according to Aristotle, were those that contained the principle of their own motion; artificial ones received this principle from elsewhere. Since the distinction was cast in terms of the principles of motion, its collapse was perhaps an inevitability, given the widespread modern rejection of the Aristotelian account of motion. However, as the passage above from Bacon emphasizes, the art/nature distinction was vulnerable on its own terms; as I will indicate, Hobbes quite deliberately attempts to undermine it. Fourth, I will examine the way Hobbes quite consciously attempts to establish the political equality of the citizens of a commonwealth by way of a subversion of the Aristotelian dictum that some are fitted by nature to rule while others are naturally fit to be ruled. Such a sense of equality is fundamental to modern political thought as we have received it, and Hobbes's own invocation of it is of interest precisely in its liminality: Hobbes does not take equality for granted; he produces an argument for it, and this argument is much more about the methodology appropriate to political philosophy than it is a claim about human nature.

Finally, I close the chapter with a discussion of Hobbes's demotion of prudence, as it relates to the apparent incapacity of prudence to ground practical reason. This putative failure is symptomatic of how Hobbes treats reasoning generally, and its bases become evident in a reading of Machiavellian *virtù* against the background of Aristotelian *phronesis*. There is both a theoretical and a doxographic reason to study Machiavelli in this regard. The doxographic reason is that Machiavelli provoked (or was symptomatic) of a more general crisis in European political thought, one with reverberations well into Hobbes's time. Machiavelli's declaration that a ruler need only appear, but not actually be, devout, proved to be enormously damaging to standard Renaissance and scholastic political thought's assumption that looking good and being good were the same thing. Machiavelli's work thus became a lightning rod for angry denunciations. It would perhaps be an exaggeration to say that the period in which Hobbes wrote was overdetermined by the need for an adequate response to Machiavelli, but I do not think that it would be a tremendous exaggeration, if the Florentine is taken as symptomatic of a general sense of crisis in sixteenth-century Europe. The theoretical reason to study Machiavelli is that, as I will argue, there is a striking sense in which Machiavelli's texts, from a modern point of view, are necessarily aporetic. They require the exercise of a faculty of judgment that is barred by the objects it studies from the epistemic resources necessary for that exercise. Viewed retrospectively, Hobbes's texts can be seen to address precisely this problem.

The remainder of the book studies Hobbes in detail. Chapter 3 examines Hobbes's forays into mathematics. This is both an obvious topic—Hobbes declares in the Epistle Dedicatory to *De Cive* that "whatever in short distinguishes the modern world from the barbarity of the past, is almost wholly the gift of Geometry" (DC *Ep. Ded.*, emphases removed)—and also a frustrating one, as almost everything Hobbes actually said about mathematics was wrong, often embarrassingly so. What are we then to make of a thinker who bases political thought on his views of mathematics? Is modern political philosophy based on a mistake? That answer is too easy: my argument is that Hobbes's mathematical writings make much better sense if we view them as governed by the assumptions of ancient Greek mathematics, rather than the emerging *episteme* of modern algebra. If I am correct, what this means in practical terms is that Hobbes actually *rejects* constructive principles at certain decisive moments in mathematics and that this rejection filters into his political thought. The apparatus of his political thought, then, will bear not only the traces of Hobbes's constructive principles but also the instabilities engendered by his incomplete adoption of them.

Chapter 4 looks at Hobbes views on language, where I will argue that the combination of his reception of scholastic nominalism and his mechanistic psychology generates a theory of signification that is almost postmodern in its rejection of any ontological foundations. Signs that Hobbes has anomalous views on language emerge both in his reduction of moral terms to linguistic ones and his reduction (in his critique of Descartes) of thought to the concatenation of words. I situate and ground these views in the context of seventeenth-century understandings of language and, in particular, the Ockhamite nominalism that Hobbes both inherits and transgresses. The sum of this investigation is the thought that Hobbes's theory of language is where his modernity is at its most radical, and that the very radicality of this theory of language is a substantial part of why Hobbes makes the move to absolute sovereignty in his political theory.

In the following two chapters, I read three central features of Hobbes's thought through the lens enabled by this somewhat unstable combination of his views of geometry and language. Chapter 5 looks at the state of nature, the constitutive feature of which is a total semiotic collapse. The state of nature, in other words, is what it is because of the collapse of stability in signification. Reading the state of nature in this way both accounts for Hobbes's terms and helps to address some lingering problems in Hobbes-interpretation, such as his discussion of contractual obligation in the natural state. Chapter 6 applies these results to Hobbes's positive political theory and looks at his distinction between the "multitude" and "people," on the one hand, and the function of the judiciary, on the other. In these cases, we can see that Hobbes's political theory, in contemporary terms, is an early version of what one might call biopolitics: politics is about the productive management of the entire field of life.

The conclusion then returns to the question with which I started: why is Hobbes of such fascination to us today? The answer in a sentence is that he is a theorist of biopolitics. How is he a theorist of biopolitics? Because his is the *poietic* thought of the productive management and development of the constitutive energy of the people. Hobbes thus definitively displaces the Aristotelian order of adapting to the natural world and the problems with the exercise of *phronesis* that Machiavelli put to that order. A second glance shows that things are not at all as tidy as this resolution suggests, for Hobbes arrives at his synthesis via two paths. On the one hand, his complete reduction of the intellectual faculty to the imagination frees language to function in all of its constitutive power, divorced from the shackles of intellectual intuition. On the other hand, his insistence on the scientific character of classical geometry grounds the original homogeneity of the state of nature and enables him to reject the emergence of construction in mathematics, as it found expression in the development of symbolic algebra. What is striking about this convergence of paths is that they pull in entirely opposite directions. The fundamental instability of the Hobbesian theory, then, gives us what I would suggest is the primary task of political philosophy today: to think biopolitics without affirming its particular result in *Leviathan*.

Chapter 2

A Genealogical Context of Modern Political Thought

Scientia propter potentia; Theorema (quod apud Geometria proprietas investigatio est) propter Problemata, id est, propter artem construendi; omnis denique speculatio, actionis vel operis alicujus gratiâ instituta est.

De Corpore *I.6*

The interpretive thesis of this book can be expressed in a phrase: for Hobbes, thought in general, and political thought in particular, is a matter of construction. The present chapter establishes some of the parameters and discursive fields within which Hobbesian thought moves, and through which my central claim can be intelligible. In that regard, it is preparatory for the more detailed analyses that follow. I will pursue several, loosely connected, threads. First, I will examine contemporary rejections of Hobbes to underscore the extent to which it was his method, and not his results, that most bothered his contemporaries. Then, I will sketch a preliminary interpretation of what I mean by the claim that Hobbes's thought is "constructive." Third, as a way of providing an initial grounding for my claim that Hobbesian thought takes politics out of the register of nature and moves it into art, I will look more generally at the erosion of the Aristotelian nature/art distinction in the seventeenth century and at evidence that Hobbes is consciously appropriating that erosion. Fourth, I look at one of the postulates enabled by the technologization of the *polis*: the thought that humans are all equal to one another. Finally, I examine Hobbes's demotion of prudence, via a discussion of Machiavelli.

2.1 Human Nature According To the Cradle, Bedlam, and a Gentleman's Kitchen

European political thought in general, and English political thought in particular, were in crisis long before Hobbes emerged on the scene.[1]

Hobbes's attempts to address this crisis were met with profound and even violent hostility. Late in his life, Hobbes felt sufficiently threatened by the Clarendon Codes that he burned many of his papers in order to reduce the possibility that he himself might be burned. He wrote various defenses of his reading of the Biblical text and began to despair of his *Leviathan* being anything more than a utopian fiction. In the English edition, he admits:

> Considering how different this Doctrine [Hobbes's] is, from the Practise of the greatest part of the world, especially of these Western parts, that have received their Morall learning from *Rome* and *Athens*; and how much depth of Morall Philosophy is required, in them that have the Administration of the Soveraign Power; I am at the point of believing this my labour, as useless, as the Common-wealth of *Plato*; For he also is of opinion that it is impossible for the disorders of State, and change of Governments by Civill Warre, ever to be taken away, till Soveraigns be Philosophers. (L 31.41, 254)

He then constructs a somewhat contrived case for optimism; by the Latin edition the case for optimism is much abbreviated, and he tersely wonders if his writing will be numerated with Plato's *Republic*, Bacon's *New Atlantis*, and "similar playthings of the mind" (OL III, 264).[2]

A full cataloguing of the response to Hobbes's work is outside the scope of this study. However, a glance at some representative responses should help to frame the extent to which, and the way in which, a Hobbesian text surprised its readers, not just for the political conclusion, but for the way it arrived at those conclusions. Robert Filmer, whose *Patriarcha* was the target of Locke's *First Treatise*, had this to say:

> With no small content I read Mr Hobbes' book *De Cive*, and his *Leviathan*, about the rights of sovereignty, which no man, that I know, hath so amply and judiciously handled. I consent with him about the rights of exercising government, but I cannot agree to his means of acquiring it. It may seem strange I should praise his building and yet mislike his foundation, but so it is. His *jus naturae* and his *regnum institutivum* will not down with me, they appear full of contradiction and impossibilities. A few short notes about them I here offer, wishing he would consider whether his building would not stand firmer upon the principles of *regnum patrimoniale*, as he calls it, both according to Scripture and reason . . . If, according to the order of nature, he had handled paternal government before that by institution, there would have been little liberty left in the subjects of the family to consent to institution of government. (1991, pp. 184–5)

The disjunct between approval of result—apparent defense of monarchy—and method, was a commonplace. Indeed, Hobbes himself baits his adversaries by advertising his methodological novelty. As noted in chapter 1, in the introductory material to *De Corpore*, and after declaring that natural philosophy began not with Aristotle but with Galileo, "the first that opened to us the gate of universal natural philosophy" (EW I.viii), he adds that, "natural philosophy is therefore young; but Civil Philosophy is yet much younger, as being no older (I say it provoked, and that my detractors may know how little they have wrought upon me) than my own book *De Cive*" (EW I.ix).

As the list of luminaries cited in this passage (Galileo, Harvey, Kepler, Gassendi, and Mersenne) makes clear, Hobbes's intellectual allegiances lie with the Mersenne circle. This allegiance was one draw of hostile fire, even as that fire incorrectly attempted to assimilate Hobbes and Descartes. The Cambridge academic John Eachard refers to Hobbes as a "Cartesian Trickster" and proposes that Hobbes's work be sent "beyond Sea: where . . . you are much read, understood, and admir'd" (1673, p. 84).[3] Similarly, Hobbes's former friend, Edward Hyde (Earl of Clarendon) says that Hobbes "thinks, that tho his Savage Country-men, and Neighbors, have yet only bin accustomed to Governments imperfect, & apt to relapse into disorders, he hath found out principles by *industrious meditation*, to make their constitution everlasting" (1676, pp. 47–8, emphasis in original).[4] Clarendon's underscored "*industrious meditation*" of course recalls the productive mathematics of the Cartesian woodstove and stands as a shorthand for all the *novatori* in general. Thus, Clarendon says later in the same text that Hobbes would "erect an engine of Government by the rules of Geometry, more infallible then Experience can ever find out" (117). Similarly, John Whitehall will accuse Hobbes of having failed to reason "from any one particular" (1680, p. 7) and proclaims with respect to one of Hobbes's arguments that "I think never any man said so before, and so with the rest of his new-found Doctrines, I pass it by" (1680, p. 87).

Eachard's *Dialogues* are of particular interest in this regard. Eachard also wrote at the intersection of popular and theological-academic culture: he was a Cambridge academic (eventually Master of St. Catharine's college and twice Vice-Chancellor of the university), but also a successful satirist, whose *The Ground and Occasions of the Contempt of the Clergy* (1670) generated both several replies and a sequel. As satires, the anti-Hobbes dialogues should be read for the picture they attempt to paint of their target, and in this regard, Eachard's naming him "Philautus" is significant, as Philautus stood both for geometry and narcissism.

First, Philolaus named an ancient Pythagorean who was reputed to have advocated a heliocentric astronomy.[5] In the dedicatory epistle to *De Corpore,* Hobbes praises Copernicus for beginning scientific astronomy, having revived the ancient opinion that the earth revolved around the sun, as expressed by Pythagoras, Aristarchus, and Philolaus (OL I, nn). The *Decameron Physiologicum* repeats the reference, first noting that few ancients—principally Plato, Aristotle and Epicurus—"seriously applied themselves to natural philosophy." Hobbes then writes that "the writings of Philolaus and many other curious students being by fire or negligence now lost: though the doctrines of Philolaus concerning the motion of the earth have been revived by Copernicus, and explained and confirmed by Galileo" (EW VII, 76). Copernicus, in turn, refers to Philolaus as being "supposed to have held that the Earth moved in a circle and wandered in some other movements and was one of the planets" (1995, I.5, p. 13).[6] There are other scattered references to Philolaus in ancient sources. When Aristotle refers negatively to the attempt to posit the generation of eternal things by the "Pythagoreans," it is likely Philolaus whose view he had in mind (*Metaphysics* 1091a13).[7] There is a direct reference in Plato: in the *Phaedo,* Philolaus is presented as having argued that suicide is not legitimate, though the details of this view seem to have eluded his pupils (61d–e).[8] Finally, Diogenes Laertius also reports a story, attributed to Hermippus, according to which Plato purchased a book by Philolaus from which he then copied the *Timaeus.*[9]

Second, seventeenth-century uses of the term suggest narcissism. For example, Spinoza opposes "philautia" to "humilitas" and says that "since this pleasure is repeated whenever a man regards his own capabilities, that is, his power of activity, the result is again that everyone is eager to tell of his exploits and to boast of his strength both of body and mind, and for this reason men bore one another" (E3P55S). In England, Francis Bacon had presented his "Of Love and Self Love" to Queen Elizabeth in honor of her accession day, November 17, 1595.[10] In it, a hermit/philosopher, soldier, and statesman, three emissaries of Philautia, the goddess of self-love, attempt to dissuade Erophilius from service to the Queen. The Squire's retorts on the latter's behalf are quite specific: to the philosopher, he demands "have not many which take themselves to be inward counsellors with Nature, proved but idle believers" (1996, p. 66)? The statesman, the "truest bondman to Philautia" whose "life is nothing but a continual acting upon a stage" is repeatedly charged with pursuing a failed attempt to "govern the wheel of fortune" (67, 66).[11]

Within Eachard's dialogue, Philautus authorizes himself with the declaration that "*Mersennus* and *Gassendus*" like "my *Book de Cive*" (1958, p. 14).[12]

The dialogue shows the epistemic distance between Philautus and Eachard's hero Timothy by invoking tropes of insanity: when Philautus insists upon the Hobbesian state of nature according to "one *demonstration* more upon thee . . . whereby it will experimentally appear, that men at first were . . . in a state of war" which will "never be utterly wash'd out till *Doomesday*" (1958, pp. 60–1), Timothy says:

> But surely it is not absolutely necessary to say all this care is taken and these defences made, because *Humane Nature* at first was, and in generall still is a *Whore*, a *Bitch*, a *Drab*, a *Cut-purse*, &c. But because there be *Doggs, Foxes, Children, Fooles, madmen, Drunkards, Thieves, Pyrats* and *Philautians*. (1958, p. 62)

At one level, then, Eachard is denying the Hobbesian account of the state of nature because the bad behaviors Hobbes lists are those of social outcasts. The passage bears a second reading, however. Since the list of irrational outcasts includes the "Philautians," that is, Hobbesians, Eachard suggests to the reader that Hobbes himself is without reason. Two consequences should be noted. First, the effect of the maneuver is to establish exactly what Eachard has said in his introduction he will do, that is, to show that the Hobbesian discourse is entirely outside the *ratio* of anything that could be called "political philosophy."[13] Second, if one reads the passage as concerning the possibility of a demonstration about human nature, then in order to preserve some sort of goodness in human nature and some sort of essential rationality within the human, Eachard has been compelled to invoke the category of the insane. That is, Hobbes's account of the human is presented not so much as wrong, but as of the irrational, the product, perhaps, of its author's deranged mind. Eachard continues:

> The short of your opinion is this, *Philautus*, that *children, fools* and *madmen*, are not very ambitious of being of the *Privy Council*; and if they were invited thereunto, would do themselves and the *Nation* but little service. So that if *right reason* (which, *Philautus*, you so much talk of, and pretend to) does determine that the *Cradle, Bedlam*, and a *Gentleman's kitchen* shall be the only standard and measure of *Humane Nature*, then truly *Philautus* must be acknowledged by all for a most mighty *Philosopher*. (1958, pp. 42–3)

Philautus denies that he says anything of the sort: he has provided a demonstration of human nature; we should note carefully that the insertion of a category of the insane is Timothy's. At the archaeological level we are observing here, madness becomes a categorical response to Hobbes's absolutism. Rather than deny that Hobbes has an adequate description of human nature, the response is to say that the demonstration applies only to the insane, who then (and therefore) have to be institutionally classified as unfit for inclusion in the political. At that point, however, the procedure is entirely Hobbesian, insofar as (as I will argue in chapters 5 and 6) the political is a process of a conceptual order's maintaining itself by excluding that which it cannot discipline.

2.2 Thinking as Construction

To say that Hobbesian thought is "constructive" or *poietic* is to say that, in a fundamental sense, it is about making things: bringing forth the objects of its knowledge. In this regard, Hobbes's view might also be considered an extreme version of what is called "maker's knowledge."[14] Documenting precisely what this claim entails will be a principal object of this study. As an initial guidepost, it might be worth establishing the difference between Hobbes's understanding of thinking and the Aristotelian schema outlined in *Nicomachean Ethics* VI and adapted, with little significant variation, by much of the scholastic tradition. Aristotle identifies the intellectual virtues as consisting in: (a) knowledge (*epistemé, scientia*) as that which studies things that are by necessity; (b) art (*techné*) as that which studies things that may or may not be and thus for things that humans produce; (c) wisdom (for the contemplation of honorable things); and (d) action (for that which changes and for which prudence would be appropriate). The fundamental distinction, as reported by scholastic commentators such as Zabarella, was between things which were necessary and those which were contingent; things which were produced could either be produced or not and therefore were specifically related to the contingent.[15] Since we cannot have *scientia* of those things which are contingent, and since things produced by our will are contingent, we cannot have *scientia* of things which are produced by our will.

In the first chapter of *De Corpore*, Hobbes outlines his mature understanding of philosophy and of ratiocination. Hobbes's understanding of necessity is complicated, and is overdetermined by his adoption of nominalist theses about the absolute power of God.[16] Strict metaphysical necessity does not exist, as to suppose that would be to suppose a limit on divine power. Physical necessity, on the other hand, is a consequence of divine omnipotence: that which happens does so by the necessity imparted by God's having freely willed it. When humans speak of necessity, in any case, they are speaking of a principle of rational intelligibility. Thus, "one cannot speak of cause, except with respect to sequence, but after the last thing, nothing follows." Accordingly, "in an action, the *principle* and the *cause* are taken to be the same thing" (DCo IX.6). Hence, "cause," taken simply, is the "aggregate of all the accidents such that, with all of them supposed, it is impossible to understand the effect as not at once produced; and, on the supposition that one of them stops, it is impossible to understand otherwise than that the effect is not produced" (DCo IX.3). The move to rational intelligibility is thus paired with an emphasis on production, and this emphasis on production offers a lens through which Hobbes's reconfiguration both of the intellectual virtues and his definition of philosophy itself becomes legible.

Hobbes explains that "philosophy is cognition, acquired by correct reason, of an effect or phenomenon from the concept of its cause or generation, and, inversely, from knowledge of the effects to the generations which are possible." Ratiocination, he adds, reduces to computation, which in turn reduces to addition and subtraction (DCo I.2). Having thus narrowly specified the operation of thought, Hobbes is able to exclude a number of things from philosophy. The subject of philosophy "is every body, for which some generation is able to be conceived, and the comparison of which to some other is able to be established by this consideration." He quickly underscores that "where there is no generation or no property, there is no philosophy" (DCo I.8). A list of exclusions from philosophy follows, of which theology is the first, since God is both simple and eternal; *Leviathan* thereby demotes "wisdom" to a form of prudence (L 3.7, 22). The next exclusion—and this is a vital point, given the importance of prudence to the Aristotelian account of ethics and thus of politics—is any sort of (either natural or political) history, which "although most useful (rather necessary) to philosophy, because such cognition is either from experience or authority, is not however ratiocination" (DCo I.8).[17]

Thus, the division of philosophy into parts according to the necessity or contingency of the objects it studies is understood according to the manner of their generation. The emphasis on generation develops over Hobbes's career, and the distance he travels can be illuminated by comparison with his earlier, more traditional definition. In the early *De Motu*, we read that "philosophy is the knowledge or theory of the general, or universal of all things in matter, in every way in which the truth is able to be demonstrated by natural reason" (DM I.1).[18] This early definition also supports a list of exclusions, whose content is similar to that in *De Corpore*, but whose justification is more Aristotelian in tone: "the end of the study of philosophy is not to influence, but to know certainly; therefore it does not consider rhetoric; and to know the necessity and truth of the consequences of a universal proposition; therefore it does not consider history: much less poetics, for these narrate singular givens, and furthermore deny the truth by profession" (DM I.3). By the time of *De Corpore*, Hobbes can write that natural philosophy studies things which "are formed by nature [*a natura rerum compaginatum*]"; civil philosophy studies bodies which "by the will of human conventions and pacts are constituted by man" (DCo I.9). In short, the growing focus on generation and production generates a redefinition of political thought along constructive lines.

As the constant exclusion of history should suggest, to determine that political thought is constructive is thus to devalue *phronesis* in particular. This is not to claim that Hobbes is alone in such a devaluation. What it is to claim is that the basis for political action will be radically transformed. As I will argue at the end of the chapter, Machiavelli's texts demonstrate the epistemic problems with prudence: the prudent individual will imitate great leaders, but the imitated leaders prove almost impossible to certify as models worthy of imitation. On Hobbesian terms, *phronesis* can generate neither secure knowledge nor a secure commonwealth because the diversity of phenomena it encounters constantly threaten to undermine the possibility of a stable foundation for knowledge and hence for the organization of a commonwealth. The generative power of nature overruns the power of thought to regulate it. The Hobbesian solution will be to extirpate the problem at the root, by fundamentally removing the *polis* from nature.

This separation of nature and the *polis* both aligns and separates Hobbes's usage of maker's knowledge from its scholastic antecedents, which (following Aristotle) took it as a given that to make something implied that one had knowledge of it. Natural objects were only known in this sense by God. Thomas Aquinas, whose views I will take as exemplary, writes that:

> In all things not generated by chance, the form must be the end of any generation whatsoever. But an agent does not act on account of the form, except in so far as the likeness of the form is in the agent, as may happen in two ways. For in some agents the form of the thing to be made pre-exists according to its natural being, as in those that act by their nature; as a man generates a man, or fire generates fire. Whereas in other agents (the form of the thing to be made pre-exists) according to intelligible being, as in those that act by the intellect; and thus the likeness of a house pre-exists in the mind of the builder. And this may be called the idea of the house, since the builder intends to build his house like to the form conceived in his mind. As then the world was not made by chance, but by

> God acting by His intellect, as will appear later (Q[46], A[1]), there must exist in the divine mind a form to the likeness of which the world was made. And in this the notion of an idea consists. (STh I^a q 15 a 1 co)

As a builder generates a house according to the idea in his intellect, so does God generate the entire world according to the idea in His intellect. To generate the whole of something requires that one possess the idea of the whole. Hence, if God created the order of the universe, he "must have the idea of the order of the universe." This idea of the order of the entire universe entails ideas of its parts, again with analogy to building. "Now there cannot be an idea of any whole, unless particular ideas are had of those parts of which the whole is made; just as a builder cannot conceive the idea of a house unless he has the idea of each of its parts." Thomas therefore concludes that "in the divine mind there are the proper ideas of all things" (STh I^a q 15 a 2 co).

When one understands something, one can pursue at least two activities: one can either contemplate the thing, or one can attempt to make it. To possess the idea of something is thus to possess either an item of speculative knowledge, an item of practical knowledge, or some combination of the two. One way that knowledge can be purely speculative arises "on the part of the things known, which are not operable by the knower; such is the knowledge of man about natural or divine things" (STh I^a q 14 a 16 co). Since natural things are not subject to human control, insofar as the *polis* is grounded in nature, humans cannot make it in the same manner as they might make a house. One might put the point this way: the matrix governing the phenomenality of natural things is external both to the phenomena themselves and to human thought, and so it serves as a limit on the ability of human ratiocination to do more than contemplate them.[19]

Although the details of this schema varied widely, that God possesses archetype ideas in some form was a sufficient commonplace that Descartes, defending himself against Hobbes, complains that he is using "the standard philosophical term used to refer to the forms of perception belonging to the divine mind" (PWD II, 127; AT VII, 181). I will return in Chapter 4 to Hobbes's engagement with Descartes and to the full radicality of Hobbes's understanding of the term "idea;" for now, I want to indicate that, even without the psychology on which his critique of Descartes is based, that is, even if one retains the broad outlines of the scholastic account of divine ideas, Hobbes has fundamentally changed the scope of practical knowledge. Hobbes's position stems from the separation of the *polis* from nature, as is evident in his realignment of the division of the arts. This realignment involves the coalescence of several terms we have been considering; according to a text of 1656, the division is as follows:

> Of arts, some are demonstrable, others indemonstrable; and demonstrable are those the construction of the subject whereof is in the power of the artist himself, who, in his demonstration, does no more but deduce the consequences of his own operation. The reason whereof is this, that the science of every subject is derived from a precognition of the causes, generation, and construction of the same- and consequently where the causes are known, there is place for demonstration, but not where the causes are to seek for. Geometry therefore is demonstrable, for the lines and figures from which we reason are drawn and described by ourselves; and civil philosophy is demonstrable, because we make the commonwealth ourselves. But because of natural bodies we know not the construction, but seek it from the effects, there lies no demonstration of what the causes be we seek for, but only of what they may be. (EW VII, 184)

The passage can be seen as a particularly clear presentation of a Hobbesian constant. As early as a 1636 letter to Cavendish, Hobbes speaks of "thinges that are not demonstrable, of wch kind is ye greatest part of Naturall Philosophy," and notes that "the most that can be atteyned unto is to have such opinions, as no certayne experience can confute, and from wch can be deduced by lawfull argumentation, no absurdity" (HC 33). The best natural philosophy, in other words, would merely be not (yet) refuted. *De Homine* (1658) outlines the same distinction, concluding that "politics and ethics (i.e., the sciences of *just* and *unjust*, of *equity* and *inequity*) can be demonstrated a priori, because we ourselves make the principles—that is, the causes of justice (namely laws and covenants)—whereby it is known what *justice* and *equity* and their opposites *injustice* and *inequity*, are" (DH 10.5).[20]

I will examine in detail to the epistemological issues posed by passages such as these, in particular the priority established of synthesis over analysis, in chapter 3. For now, I would like to emphasize two points. First, note that "man,"[21] considered politically, is not necessarily determined by nature, which is to say that if "man" is *initially* or immediately determined by nature, that natural determination can nonetheless be qualitatively changed by application of the principles of political philosophy. Whatever the "natural" determination of the human is will therefore become irrelevant to political thought except as a meaningless residuum left behind. In particular, the evacuation of natural meaning is the evacuation of intelligible natural difference; such evacuation means that people are equal *de jure*.

Second, the emphasis on generation is joined to an emphasis on language. Hobbes adds to the passage from *De Homine* that "for before covenants and laws were drawn up, neither justice nor injustice, neither public good nor public evil, was natural among men any more than it was among beasts" (DH 10.5). In *Leviathan*, this point grounds a brief outline of how not to proceed in political philosophy. "Ignorance of the signification of words," he writes, "disposeth men to take on trust, not onely the truth they know not; but also the errors; and which is more, the non-sense of them they trust. For neither Error, nor non-sense, can without a perfect understanding of words be detected" (L 11.18, 73). From this deficiency proceeds others: different names applied to the same thing, as for example calling something "opinion" if one likes it but "heresy" if one does not (L 11.19, 73).[22] Worse still, ignorance of the "causes, and originall constitution of Right, Equity, Law, and Justice, disposeth a man to make Custome and Example the rule of his actions." Such a person then vacillates between the claims of custom and reason, according to whichever serves his apparent interest at the moment. He concludes that this is:

> The cause, that the doctrine of Right and Wrong, is perpetually disputed, both by the Pen and the Sword: Whereas the doctrine of Lines, and Figures, is not so; because men care not, in that subject what be truth, as a thing that crosses no mans ambition, profit, or lust. For I doubt not, but if it had been a thing contrary to any mans right of dominion, or to the interest of men that have dominion, *that the three Angles of a Triangle, should be equall to two Angles of a Square*; that doctrine should have been, if not disputed, yet by the burning of all books of Geometry, suppressed, as farre as he whom it concerned was able. (L 11.21, 74)

Thus the Hobbesian program must attend to the correct specification of its principles, in particular on the basis of geometry and an understanding of language; and it must also confront the problem of how to institutionalize itself in order to manage the affects.

2.3 The Monster of Malmesbury

If I am right that, for Hobbes, politics is separated from nature and displaced into the realm of artifacts, understanding this move will require understanding what conceptual structures were available for distinguishing natural and artificial objects. The relevant scholastic authority was Aristotle, according to whom, of "things which are generated, some are generated by nature, others by art, still others by chance; but every thing generated is generated by something, and out of something, and it becomes something" (*Metaphysics* 1032a13–14). As I will discuss in more detail below, Aristotle is not particularly interested in generation by chance. Indeed, as the passage from St. Thomas discussed in the preceding section indicates, subsequent interest focused on the way that things were generated and the status of the formal cause in that generation: for things generated according to nature, the form would be somehow spoken of according to "natural being," whereas for artifacts, the form would be attributed to "intelligible being." In Aristotle, the fundamental distinction had to do with how one might conceptualize the way that the object in question moves, and in particular, where the principle of its motion is located.

Hence, in *Physics* II, he conceptualizes the distinction between things that are by nature and those that are by art. Examples of the former "appear to have in themselves a principle of motion and of standstill," whereas an example of the latter has "no natural tendency in itself for changing" (192b14–15). Surrounding text indicates that Aristotle's primary concern is less with the nature/art distinction *per se* than in deploying it to prove that nature is teleological. His point appears to be that if one wants to say that art imitates nature, then since art is clearly teleological, this imitation is only possible on the assumption that nature is also teleological.[23] As I will indicate, the early modern rejection of teleology is intricately bound with the rejection of the nature/art distinction; Hobbes will retain the importance of causality in understanding art but will drop any pretext that art must imitate nature.

According to Aristotle's *Physics*, the external motion imparted to things by art can be understood in at least two ways: "there are two arts which rule over matter and have knowledge of it—the art which is concerned with the use of it and the art which directs the production of it" (194b1). Both use and production are teleological in the sense that they occur for the sake of something, and Aristotle rephrases the original nature/art distinction in teleological terms: "now in objects produced according to art, it is we who produce the matter for the sake of some function, but in natural objects it is there all along" (194b7). This sense of teleology inherent in natural objects opens a two-pronged inquiry into the relation between teleology and necessity that Aristotle addresses in *Physics* II.8, namely, "why nature is a cause for the sake of something" and "how necessity exists in physical things" (198b10–11). One may phrase the relation in the form of a question: what prevents natural things from moving by necessity, and not teleologically? Aristotle answers by deploying the category of chance across a pair of disjunctions. On the one hand, things happen either always or for the most part, or by chance. On the other hand, things occur either by chance or for the sake of something. Since natural things occur always or for the most part, they cannot occur by chance. Since they cannot occur by chance, they occur for the sake of something. So natural things occur for the sake of something.

That the early moderns launched a wholesale attack on scholastic developments of this account of nature and account of natural motion derived from it is well-known. Where the early moderns both sharply curtailed or even eliminated the role of teleological explanation in physics and (therefore) tended to reduce all motion to local motion in the sense of a change in place, the Aristotelian and scholastic usage of teleology enabled a number of qualitatively distinct ways of understanding motion. The complexity of these distinctions, and their propensity to bleed into other issues, can be exemplified in the Coimbra edition's gloss on the nature/art passage:

> He [Aristotle] establishes the distinction between those things which exist by nature, and which are through other causes, especially artifacts: for those things which exist by nature all contain the principle of motion and rest innately [*insitum*], some indeed motion to a place, as celestial bodies; others by alteration, as other bodies; or also growth and decay, as living things; and truly, which are through other causes, as beds, clothes and other things worked out by the skill of art, such things which do not have in themselves the cause of motion or of rest, unless by accident, because indeed it happens that the bed is of silver, or wood, or consists of some other similar matter.[24]

Thus, the umbrella term "motion" in Aristotle comes to be defined broadly as "the actuality of the potentially existing qua existing potentially."[25] Early modern efforts to break with this definition tended to emphasize its alleged obscurity and to offer metaphysically streamlined alternatives. Thus, Descartes famously quips that "these words are so obscure that I am compelled to leave them in Latin because I cannot interpret them" (PWD I, 94; AT II, 39). Hobbes reduces all motion to local: "motion is the continual abandonment of one place and acquisition of another."[26] Spinoza emphasizes the connection with teleology and announces that all of the prejudices arranged against his *Ethics* "hang on this one, that men evidently commonly suppose that all natural things, as themselves, act according to an end" (E1Appx).

Partly because it involved teleology, and although philosophers such as Descartes talked about it less, the distinction between nature and art was also under pressure by the seventeenth century. Here the root of the problems lay in the instability of the explanatory priority accorded to nature over art. Although in Aristotle, the distinction was arguably perspectival or epistemic, by the Middle Ages, art had been accorded clear second-citizen status: nature was fundamental; artifacts imitated nature. The attack on natural teleology left this explanatory structure open to the worry that the Aristotelian schema had borrowed concepts from art to explain nature, and not the other way around.[27] Descartes' reduction of animals to machines in the *Discourse* is well known; by the end of the century, Locke was able precisely to adopt schema for understanding nature from art, explaining the identity of an animal in terms of a watch (*Essay* 2.27.5). Earlier evidence of slippage in the priority of nature came from several fronts. Again, I will use the Coimbra *Physics* commentary as exemplary. The sixth question following the text and glosses on *Physics* II.1 amounts to a cataloguing of strange artifacts and other automata, and the commentators note parenthetically that "we omit very many others of this type" (*In Phys* 2c1Q6a1, 217). Thus, the first argument catalogues such apparently self-moving artifacts as the statues of Daedalus, about which Plato reports that "if no one ties them down, [they] run away and escape" (*Meno* 97d); and the tripods recounted in Flavius Philostratus' *Life of Apollonius of Tyana*, which "stepped forth like those of the Pythian Temple, but of their own accord" (III.27).[28] The early modern period indeed saw a profusion of such marvels, and they provoked anxiety across any number of cultural registers. By 1612, for example, Rudolf II had assembled a huge inventory of them, and when Prague was sacked in 1620, the inventory was scattered across Europe.[29]

The second argument recounts astronomical events and entities which appear to produce effects that do not pertain to their nature, and the third refers to the work of the alchemists. This work "does not seem to deny that many works of nature are effected by art, but that in several things art does not emulate nature, but conquers it" (*In Phys.* 2c1q6a1, 218). Alchemy had been a lightning rod for discussions of the Aristotelian art/nature relation and its transgression since the thirteenth century, when Avicenna's *De conglegatione* found its way into Aristotle's *Meteorologica*. Avicenna had denounced alchemy on the grounds that art could only imitate nature, which had metaphysical priority; and that alchemists could not, in any case, be transmuting metal, since they did not know the relevant species-determining characteristics.[30] The Coimbra commentator's list thus presents us with three kinds of motion that move in a way that cannot be reconciled with the axioms that natural motion is teleological and that only natural things contain the principle of their own motion. The first and third kinds, in particular, seemed to present instances where art subverted natural motions and bent them to its own ends.

The commentators' resolution to the problem they pose is to emphasize that the art does not bring about the motion of the artifact but modifies it; hence the artifact is moved "not by its own form, not by art, but by nature, lying in an orderly way by art in the machine." They conclude, perhaps too quickly, that "thus art does not bring about these motions, but modifies them and adjusts them" (*In Phys.* 2c1q7a2, 222). The conclusion, in other words, amounts to a restatement of principle: motion inheres in natural things; artificial things are composed of natural things, so the motion of artificial things is the result of the motion of the natural things which compose them. But there is a problem here; if art "modifies" or "adjusts" natural motion, in what sense does it not provide the "principle" of that motion, particularly given that art seems actively to interfere with the course that nature would otherwise follow, and not just draw it out or help it along? How does one draw the necessary distinction in a nonarbitrary way? One way to focus the question would be to look at the way that the principle of motion is said to inhere in natural objects. How, in other words, does this natural motion inhere in the natural object, and why can it not be inside the artificial one, *qua* artificial object? And what about the results of the alchemists? The commentators suggest that "if alchemy makes gold, it does not bring it about by its own force, but by applying active natural things to passive ones" (*In Phys.* 2c1q7a2, 222). This distinction has the merit of solving the immediate difficulty, but it seems rather hard to cabin: could not any motion whatsoever be understood in analogous terms?

Evidence of the permeability of the nature/art boundary, if not its colonization by the productive arts, is present elsewhere even in Aristotle's own text. One finds there not automata but another seventeenth-century cultural obsession, monsters, though Aristotle reassures us that such aberrations as the "man-faced offspring of oxen mentioned by Empedocles perished and still do so" (*Physics* 198b32). Since monsters might again be evidence of nonpurposive natural activity, a certain finesse is again required. The Coimbra commentators therefore suggest that a monster is "rightly a natural effect, and only degenerate in terms of the disposition of the species." Why is it a natural effect? By analogy to art; we know that in the case of "artificial things, although they are directed by the rule of art, monsters are seen; these however, not properly, but only by analogy to natural things, are customarily called monsters." Therefore, the degeneracy is specific and explicable because of the intervention of an "error" (*In Phys* 2c9q5a1). With monsters thus banished to the realm of mistakes, Aristotle turns to things that happen as they should. Here, again, analogies traffic across the nature/art border, making it difficult to say if the understanding of art is modeled on nature or vice versa. In adducing an example of the point that all of the stages of activity toward an end are themselves done for the sake of that end, Aristotle suggests that "if a house were generated by nature, it would have been generated in a way similar to that in which it is now generated by art." He follows that "in general, in some cases art completes what nature cannot carry out to an end, in others, it imitates nature" (199a15). That art imitates nature, however, is then taken as evidence that things according to nature are also purposive; it therefore remains unclear what the epistemic status of mimesis is in this context.

Monsters of this sort also make an early appearance in Hobbes. In *Elements of Law,* he notes that "upon the occasion of some strange and deformed birth, it shall not be decided by Aristotle, or the philosophers, whether the same be a man or no, but by the laws" (EL 29.8).[31] The passage not only illustrates the sense in which interruptions of natural motion were of sufficient import to make their way into a discussion of sovereign powers, but also the extent to which the Hobbesian emphasis on construction changes the discursive stakes of apparent interruptions of natural motion. Hobbes's position is that what counts as "human" is a matter for legal and political determination, not natural. In this sense, the category of the human itself becomes artifactual. More precisely, the ontological question of whether something is or is not human by nature is declared strictly irrelevant.

The preceding passages should be taken as cartographic indicia of one of the ways in which tensions within and upon late scholastic physics played out. What is less well-noted than the tensions within physics is that those tensions are replicated in political thought. This replication should nonetheless be unsurprising, as Aristotle's political thought moves within the same triad of purposiveness, nature, and art that his physics does. For example, in the passage where he refers to the Daedalus statues, Aristotle distinguishes between animate and inanimate instruments; one example he adduces is that of a slave, which "is an animate possession, and every servant is like an instrument in charge of other instruments" (*Politics* I.4, 1253b34). If every instrument could perform on its own, like the statues of Daedalus or the Hepheaestean tripods, then "a master-artist would need no servants and a master would need no slaves" (1254a1). A political implication of the erasure of the nature and art distinction, then, is a change in the way relationships of master and servant can be thought. Since Aristotle will also—and I will return to this in much greater detail—declare that some people are "slaves by nature," there is even a sense in which his own categorical boundaries tend to blur on this point: to be a slave is to be an instrument, but to be an instrument is to be by art.

There are several ways that a blurring of the nature/art distinction along these lines could have political effects. On the one hand, if a slave is an animate instrument, and if the conceptual distinctions that preserve a realm of "nature" from that of other objects are eroded, then the problems of political philosophy should become less about identifying a naturally ruling class and should focus instead on questions about how to decide the criteria relevant for establishing rule. Perversely, we all become slaves by nature in the sense that the *polis* becomes open to the sorts of social engineering projects that had previously been barred by nature. Somewhat less dispiriting, perhaps, one can recognize in such a move an initial formulation of the thought that all people are equal. As I show in the next section, Hobbes will take the artifactual status of politics in precisely this direction. So too, the move to natural equality allows the question of democracy to be posed quite radically, since it is no longer the case that the *demos*' being unfit to rule can be attributed to nature. As Aristotle had already indicated, if instruments could perform on their own, neither servants nor slaves would be necessary.

The nature/art distinction also recurs with political implications in Aristotle's *Politics* discussion of teleology. There, the definition of politics incorporates a by-now-familiar subordination of activities to their ends. The *polis*, he says, starts with a "union of those who cannot exist without each other, that is, a union of male and female for the sake of procreation." He underscores that this motion is by nature, noting that "the tendency in men, as in the other animals and in plants, to leave behind their own kind is natural and not the result of deliberate choice" (1252a28–30). This movement of association, begun in natural dependency, culminates in the *polis*. He writes:

> A complete association composed of many villages is a *polis*, an association which (a) has reached the limit of every self-sufficiency, so to speak, (b) was formed for the sake of living, but (c) exists for the sake of living well. For this reason, every state exists by nature, if indeed the first association too existed by nature; for the latter associations have the state as their end, and nature is the end. (1252b28–32)

It is in this sense that "man is the political animal," since human beings naturally seek out political associations for their own preservation. At the same time, the political art serves an important role in bringing this natural movement to successful fruition; even if it is not necessary for the occasion of procreation and living, it is necessary for "the association which aims in the highest degree and at the supreme good" (1252a7). The political art is executed by the lawgiver, and the foundation of this claim ultimately traces to Aristotle's argument that "that which can foresee by thought [*dianoia*] is by nature a ruler or by nature a master" (1252a32). Although the details of the relation between nature and art are (as always) somewhat complex, Aristotle's general point is to suggest that the constructive practice of the political art is a necessary component to the fulfillment of human nature. Thus, to live in a just *polis* is the greatest human good; and to be separated from it is to undergo the greatest possible corruption, "just as man when perfected is the best of animals, so he is the worst of all when separated from law and judgment" (1253a32).[32]

Against this background, both Hobbes's rhetoric and his argument structure appear to be precisely and deliberately placed. From its titular reference to the Biblical sea monster, *Leviathan* presents itself as a monstrous text, and its first words engage in a thorough deconstruction of the nature/art distinction as it applies to the *polis*:

> NATURE (the Art whereby God hath made and governes the World) is by the *Art* of man, as in many other things, so in this also imitated, that it can make an Artificial Animal. For seeing life is but a motion of Limbs, the beginning whereof is in some principall part within; why may we not say, that all *Automata* (Engines that move themselves by springs and wheeles as doth a watch) have an artificiall life? For what is the *Heart*, but a *Spring*; and the *Nerves*, but so many *Strings*; and the *Joynts*, but so many *Wheeles*, giving motion to the whole Body, such as was intended by the Artificer? *Art* goes yet further, imitating that Rationall and most excellent worke of Nature, *Man*. For by Art is created that great LEVIATHAN called a COMMON-WEALTH, or STATE, (in latine CIVITAS) which is but an Artificiall Man. (L Intro, 9)[33]

The later Latin edition's engagement with Aristotelian terms is even more striking, and deploys the Aristotelian categories to arrive at a distinctly anti-peripatetic conclusion. "For since life is nothing other than an art of motion, the principle of which is internal to the principal part of some body, what indeed prevents us from saying that every automaton . . . also has artificial life?" He concludes: "the great Leviathan, which is called the *civitas*, is a work of art and an artificial man, although reasoned out by natural man for the sake of his protection and welfare" (OL III, 1).

Hobbes's reasoning is utterly foreign to the Aristotelian framework against which it is directed. For Hobbes, since living components of nature are in motion, and since natural motion occurs through the coordinated motion of parts, then the coordinated motion of parts can be considered as alive. Hence art can imitate, even create, (artificial) life, and the state should be considered as precisely such an artificial life. Even more strangely, from an Aristotelian point of view, the constitutive parts of the commonwealth do not, by themselves, tend toward the coordinated motion of the commonwealth. Left in the state of nature, human beings would devolve into a violent *bellum omnium in omnes*. In other words, a transformative motion is required, but it is not one that can be understood as the completion of a natural motion. The procreative natural motion is not by nature on the way to a commonwealth. Rather, it seems that art is acting contrary to nature in inducing a motion that would not naturally occur: art acts *praeter naturam*. Hobbes thus opens *De Cive* by affirming two maxims whose *prima facie* incompatibility emphasizes the caesura that civil philosophy must bridge: "man is a God to man and man is a wolf to man" (DC *Ep. Ded.*).[34]

At this point, and as one would expect, Aristotle loses interest; he is dismissive of the need to inquire further into those things which happen contrary to nature, as they can simply be said to happen by chance; the seventeenth-century English edition of his *Rhetoric* silently adds the example of "monsters, etc."[35] Understanding things by chance thus implicates questions about the possibility of understanding phenomenality in general, particularly for phenomena (such as monsters or historical events), the regularity of which is either not present or not evident. Hobbes's mature writings reflect this connection. Of natural phenomenality, Hobbes is able to say as late as *De Corpore* that "*to fainesthai* [appearing] itself" is the "most admirable" of natural phenomena (DCo XXV.1).

More strikingly, Hobbes's interest in the category of chance occurs in his political thought, and precisely in the context of how Aristotelian derivations of political philosophy are fundamentally misguided. Early in *De Cive*, he remarks:

> The majority of previous writers on public affairs either assume or seek to prove or simply assert that Man is an animal born fit for Society,—in the Greek phrase, *zoon politikon*. . . . This Axiom, though very widely accepted, is nevertheless false; the error proceeds from a superficial view of human nature. Closer observation of the causes why men seek each other's company and enjoy associating with each other, will easily reach the conclusion that it does not happen because by nature it could not be otherwise, but by chance [*quod aliter fieri naturâ non possit, sed ex accidente*]. (DC I.2)

Pressed by readers of the first edition of the book, Hobbes adds a note to the second that, given the universality of human society, "it may seem a claim of weird foolishness" to claim that "man is not born fit for society." However, he digs in his heels: civil societies are not mere gatherings; they are federations, and those require reason. Infants and those who lack training are unfit for society, and even those adults who desire society may not be fit to enter it. Therefore, he concludes, "man is made fit for society not by nature, but by training" (DC I.2, note). This final claim amounts to an explicit repudiation of the Aristotelian dictum that "the political art does not make men but takes them from nature and uses them" (*Politics* 1258a22).[36] Indeed, Hobbes had said as much even earlier, in the *Elements of Law,* distinguishing human society from the natural sociability of bees, he concludes that "natural concord, such as is amongst these creatures, is the work of God by the way of nature; but concord amongst men is artificial, and by way of covenant" (EL 19.6). The result from Aristotelian point of view is that the terms of politics are no longer governed by either art or nature but according to a third category, initially posed in the Renaissance in the terms of a rediscovered pseudo-Aristotelian text: mechanics.[37] The operative terms of politics are derived not from physics or from theology but, as Hobbes puts it, from "nothing but human wit" (EL 20.1).

2.4 An *Ethos* of Equality

As the above should begin to demonstrate, much of what is at stake in the reception of Hobbes has to do with how he represents the "natural" human condition. What is important is not that Hobbes posits an original condition for humankind—that move is a commonplace. What is peculiar is the way in which he does so; in particular, the Hobbesian state of nature does not answer to any actual condition, either historical or Biblical. Indeed, it cannot do so, because Hobbes claims that it is a condition of equality.[38] I will return in more detail to the Hobbesian state of nature; in this section, I will focus on Hobbes's posit of natural human equality not only as a move against Aristotle, but also as responsive to concerns about pedagogic and theoretical method that were widely debated in late scholastic texts.

In all of his major political works, Hobbes makes it quite clear that the equality of the state of nature is a methodological principle. That is, Hobbes never says that people *are* equal in the state of nature—he says it would produce better political philosophy if they were taken *as* equal. Hence, in the early *Elements of Law,* he notes that because the weak can easily kill the strong, "we may conclude that men considered in mere nature, ought to admit among themselves equality; and that he that claimeth no more, may be esteemed moderate" (EL 14.2). Although in this early writing the methodological premise has a loose ontological basis, the text still includes several references to discourse and ratiocination: "moderate," "ought to admit," "esteemed," and "claimeth no more." The state of nature is not, or not simply, an ontological postulate but a requirement at once grammatical and ethical. It is grammatical in the sense that the specification of the state of nature is part of the logical space in which political philosophy is intelligible. It is an ethical requirement in the sense that it determines the appropriate *ethos* for engaging in political theory. That *ethos* is one whereby people treat the state of nature *as if* its inhabitants were all one and the same.

An inversion of Aristotle is immediately apparent in Hobbes's "moderate": an Aristotelian mean is found by a process of triangulation, between the excess and deficiency of any given activity. For Hobbes, the mean is indeed generated with reference to an activity, but after that, it applies universally. First, one establishes the mean; then one applies it. A thinker such as Grotius had already indicated problems with the Aristotelian mean, suggesting that it was "not without reason [that] some of the Platonists and early Christians" departed from it, as the principle of moderation "led him to unite distinct virtues . . . into one; to assign to truth extremes between which, on any fair premiss, there is no possible coordination . . . and to apply the designation of vice to certain things which either do not exist, or are not in themselves vices" (1962, Prol. §43). Still, "among the philosophers Aristotle deservedly holds the foremost place, whether you take into account his order of treatment, or the subtlety of his distinctions, or the weight of his reasons" (1962, Prol. §42). The eclectic Grotius, furthermore, praises precisely the ethos of the scholastics for its moderation and for the way in which scholastic disputations are exemplary of peaceful practices:

> When the Schoolmen agree on a point of morals, it rarely happens that they are wrong, since they are especially keen in seeing what may be open to criticism in the statements of others. And yet in the very ardour of their defence of themselves against opposing views, they furnish a praiseworthy example of moderation; they contend with one another by means of arguments—not, in accordance with the practice which has lately begun to disgrace the calling of letters, with personal abuse, base offspring of a spirit lacking self-mastery. (1962, Prol. §52)[39]

Hobbes is opposed to this view on almost all points, reserving for the geometers the ethological praise here accorded to the schoolmen and heaping constant scorn on the disputational practices of the latter.

The parallel passage in *De Cive* about equality is also quite striking. There, Hobbes writes:

> If then men are equal by nature, we must recognize their equality; if they are unequal, since they will struggle for power [*imperio*], *the pursuit of peace* requires *that they be regarded as equal.* And therefore the eighth precept of natural law is: *that everyone should be considered equal to everybody else by nature [ut unusquique naturâ unicuique aequalis habeatur]*. Contrary to this law is called PRIDE. (DC III.13; emphasis in original)

The ontological basis in the ability to kill and be killed, though still the basis of natural equality in *De Cive* I.3 (which rehearses the discussion of weakness and the ability to be killed), has receded substantially in the text immediately surrounding the above passage, which refers hypothetically to contests of the wise and the strong. The formulation in *Leviathan* is likewise explicit about the methodological question it raises. "The question who is the better man, has no place in the condition of meer nature," he writes, concluding again a few sentences later that:

> If Nature therefore have made men equall; that equalitie is to be acknowledged: or if Nature have made men unequall; yet because men that think themselves equall, will not enter into conditions of Peace, but upon Equall terms, such equalitie must be admitted. And therefore for the ninth law of Nature, I put this, *That every man acknowledge other for his Equally by Nature.* The breach of this Precept is *Pride.* (L 15, 107)[40]

The equality of the state of nature, then, has nothing to do with nature at all: it is instead the product of specifically bracketing any detailed consideration of nature itself.

Having thus laid down this methodological principle, Hobbes attempts a reversal of the Aristotelian schema, as he himself says in the paragraph in which the law in question is derived. As suggested in the previous section, Hobbes is specifically engaged in negating the proposition that "the political art does not make men but takes them from nature and uses them" (*Politics* 1258a22), in these passages via a negation of the Aristotelian declaration that some are slaves by nature.[41] For Hobbes, the political art will be precisely about making things. Hobbes thus glosses Aristotle's passage declaring that some people are slaves by nature: "he putteth so much difference between the powers of men by nature, that he doubteth not to set down, as the ground of his politics, that some men are by nature worthy to govern, and others by nature ought to serve" (EL 17.1). In *Leviathan*, he adds that this is done "as if Master and Servant were not introduced by consent of men, but by difference of Wit: which is not only against reason, but also against experience." The reason is simple: "there are very few so foolish, that had not rather governe themselves, than be governed by others" (L 15.21, 107 cf. the almost identical language in DC III.13). For Hobbes, such deference to natural difference is a bad foundation, for both a theoretical and an ethical reason. Hence, it "hath not only weakened the whole frame of his politics, but hath also given men colour and pretences, whereby to disturb and hinder the peace of one another" (EL 17.1).

It is a principal object of this book to understand some of the theoretical apparatus behind Hobbes's rejection of Aristotle; here I want to briefly consider the ethical considerations behind that rejection, as a cluster of related issues emerges out of scholastic tradition. In brief, the scholastic claim is that theory has consequences for those who hear it, and so any question about the correct exposition of theory is in part an ethical question. That an auditor could be damaged by inappropriate material was a medieval commonplace. Maimonides, for example, includes very precise instructions for the study of his *Guide*. It is for "one who has philosophized and has knowledge of the true sciences, but believes at the same time in the matters pertaining to the Law and is perplexed as to their meaning because of the uncertain terms and the parables" (I, 10). Furthermore, "those who are confused and whose brains have been polluted by false opinions and misleading ways deemed by them to be true sciences . . . will flee from many of its chapters" (I, 16). Similarly, St. Thomas opens his *Summa Theologiae* by explaining that it will "treat of whatever belongs to the Christian religion, in such a way as may tend to the instruction of beginners." He then immediately adds that "we have considered that students in this doctrine have not seldom been hampered by what they have found written by other authors" (STh pro).

Improper instruction, then, renders one unable to apprehend the truth. In defending philosophy against the charge that it destroyed faith, Averröes suggests that "the damage done to people by demonstrative books is lighter, because for the most part only persons of superior natural intelligence become acquainted with demonstrative books, and this class of persons is only misled through lack of practical virtue, unorganized reading, and tackling them without a teacher" (1963, p. 178). He warns, however, that when interpretations of the law are "expressed to anyone unfit to receive them—especially demonstrative interpretations because of their remoteness from common knowledge—both he who expresses it and he to whom it is expressed are led into unbelief" (1963, p. 181). In short, a robust tradition leads one to expect precisely the blend of pedagogy and theory expressed in the Hobbesian passage in *Elements of Law*, we could similarly expect Hobbes to pay great attention to the order of his exposition, the preparedness of his readers, and to the material consequences of incorrect theory. If one substitutes the desire to obey one's civil magistrates in Hobbes, for faith (understood as orthopraxy, correct adherence to law) in the Averroistic account, the parallel is striking:

> And by reading of these Greek, and Latine Authors, men from their childhood have gotten a habit (under a falseshew of Liberty,) of favouring tumults, and of licentious controlling the actions of their Soveraigns; and again of controlling those controllers, with the effusion of so much blood; as I think I may truly say, there was never anything so deerly bought, as these Western parts have bought the learning of the Greek and Latine tongues. (L 21.9, 150)[42]

The authors Hobbes cites are the mainstays of the scholastic and humanist traditions, Aristotle and Cicero.

I do not wish here to resolve all of the questions raised by the lineage suggested, but I will gesture to a moment in Aristotle which provides a framework in which seventeenth-century discussions of method can be situated.[43] In the *Topics*, Aristotle argues that "against a young man you should apply your training in inductive methods, against an expert your training in deductive methods" (164a12). This is because "induction is more convincing and clear and more easily grasped by sense-perception and is shared by the majority of people, but reasoning [*syllogismos*] is more cogent and more efficacious against argumentative opponents" (1105a17–19; cf. also *Topics* VIII.2). Early modern responses to the questions implicit in this formulation varied widely.[44] Hobbes, as I will show in Chapter 4, is barred by theoretical considerations from attaching too much weight to analysis. However, even a cursory reading of the English *Leviathan* in particular suggests that he is offering images which appeal to sense perception, particularly in his descriptions of the devolution of societies into warfare (for a full discussion, see ch. 5.1).

When Aristotle considers the instructional exposition of political theory, he concludes that "a young man is not a proper student of politics" for two reasons. First, he is "inexperienced in actions concerned with human life"; and, second, he is "disposed to follow his passions, [so] he will listen in vain and without benefit, since the end of such discussions is not knowledge but action" (1095a2–5; cf. *Rhetoric* II.12–14). The basis for such comment is, of course, natural difference, and so when Hobbes claims that the invocation of natural difference itself is the problem, he requires a new ground which satisfies both pedagogical and theoretical concerns. Such a ground is definition and the clarification of key terms. As he puts it in *Leviathan*, "if the first ground of such Discourse [any *scientia*], be not Definitions . . . then the End or Conclusion, is again OPINION, namely of the truth of somewhat said, though sometimes in absurd and senseless words, without possibility of being understood" (L 7.4, 47). All of Hobbes's major writings thus begin with definitions and the clarification of terms.

Hobbes's frequent polemics against the scholastics on the point notwithstanding, there was also strong medieval precedent for beginning appropriate instruction with the clarification of terms. Hence, Porphyry's *Isagoge*, which became the standard introductory text in the scholastic curriculum, is itself a clarification of metaphysical terms such as genus, species, difference, and property. Porphyry underscores that his task is introductory: "I shall attempt, in making you a concise exposition, to rehearse, briefly and as in the manner of an introduction, what the older masters say, avoiding deeper inquiries and aiming suitably at the more simple" (2003, Pfc., 3). However, he also gestures to the more complex topic implicit in the definitional task he sets himself, the status of the terms being defined. Hence, on account of the difficulty in the question, Porphyry will demur "about genera and species—whether they subsist, whether they actually depend on bare thoughts alone, whether if they actually subsist they are bodies incorporeal and whether they are separable or are in perceptible items and subsist about them" (*ibid.*). His later commentators did not fail to draw attention to the issue. To cite only three examples, Averröes emphasizes Porphyry's restriction of his discussion to logic: "we will treat of these things only insofar as they are logical" (*In Porphyrium*, 1.i). Gersonides points to the underlying philosophical problem, provides a brief survey of the problem of universals (basing his discussion on Plato and Aristotle), and suggests that this question is logically prior to the definition of terms: "at first we will consider by what agreement many individual things ascend to one name [*quo pacto ascendant multa individua ad unam nomen*], and when we have one name and when several" (1562, 2a). The Coimbra commentators feel compelled to devote four articles to the topic, including a detailed refutation of the nominalists, citing Ockham as a prominent example (2001, pp. 80ff.).

The need for appropriate pedagogy, then, led to questions of the meaning of terms and questions about the meaning of terms led naturally to questions about the status of those definitions. As Gersonides' reference to *pactum* suggests, the question about agreement on these topics was also present, implicitly at least, especially among those who denied accessible, extramental existence to universals. In other words, questions of textual presentation, analysis, synthesis, method, and language were intimately tied in the scholastic tradition with and against which Hobbes worked. I will explore and attempt to disentangle some of these issues in subsequent chapters. As I will argue, Hobbes's own discussion is heavily indebted to that in Ockham, even as he departs from it considerably. As Hobbes has to discuss universals, so does he also have to discuss method. The Hobbesian trope for his understanding of method is "geometry," but, as I will show, what Hobbes means by geometry is not self-evident, and his invocation of geometry is deeply imbedded in his understanding of the relation between analysis and synthesis.

In the context of his rejection of Aristotle's "slaves by nature" and consideration of natural difference, the important point, as I have emphasized, is the caesura established between the natural and the political. Hobbes says exactly this: "the question, which is the better man, is determinable only in the estate of government and policy, though it be mistaken for a question of nature" (EL 17.1). What might at first appear as a simple reversal—from Aristotelian inequality to Hobbesian equality—becomes on closer examination the displacement of this ontological consideration onto questions of method. Whether or not those people actually are equal disappears as irrelevant, a question whose answer is best not attempted. The rejection of the *zoon politikon*, in other words, is for Hobbes deeply rooted in a complex series of issues whose provenance is found not just in questions of method *per se* but in the status of the language in which those issues are expressed.

2.5 Corrupting *Phronesis*

As I have suggested, one of Hobbes's moves is the attempt to make political philosophy a *scientia*; this move has as its corollary the demotion of prudence. Renaissance discussions of prudence in the context of politics had two natural referents: Aristotle's discussion of *phronesis* in the *Nicomachean Ethics*, and Machiavelli's discussions of *virtù*, especially in *The Prince*.[45] In what follows, I want to treat Machiavelli's *virtù* as an interpretation of *phronesis* that denies it any substantive notion of one's ends. Doing so allows one to frame Machiavelli as attempting to sustain an account of practical reason as prudence (as opposed to *scientia*), while maintaining a thoroughly materialist understanding of power, an understanding for which power is substantially an exercise in semiotics. As I detail in Chapter 4, Hobbes ends up with this sort of understanding on metaphysical grounds, and he views the lack of an eidetic *summum bonum* as fatal to the viability of prudence. For Hobbes, Machiavelli not only presents more evidence of the failure of pre-Hobbesian political philosophy but also, in the denial of a *summum bonum*, articulates an important truth about power, one that Hobbesian theory attempts to incorporate.

Early in *Leviathan*, we read that there is nothing specifically human about prudence, and that "there be beasts, that at a year old observe more, and pursue that which is for their own good, more prudently, than a child can do at ten" (L 3.9, 28). More generally, the excessively prudent individual, Hobbes adds, becomes captive to events which are out of his control, and, like Prometheus, "hath his heart all the day long gnawed on by fear of death, poverty, or other calamity, and has no repose, nor pause of his anxiety, but in sleep" (L 12.5, 76). The difficulty is a theoretical one, grounded in the general point that "signes of prudence are all uncertain; because to observe by experience, and remember all circumstances that may alter the successe, is impossible."[46] As I will illustrate in detail in the next chapter, this remark amounts to a précis of Hobbes's critique of analysis as a method of inquiry. Here, there is an additional epistemic complexity, suggested by his subsequent reference to "reading . . . Politiques and History" (L 5.22, 37) as exemplary of efforts at political prudence. The problem is that the interpretation of events is mediated by the historical record as presented in texts, and thus the events themselves are ultimately irretrievable. One can only interpret and comment, but then one is stuck in an epistemic dead-end analogous to the one Hobbes identifies with legal interpretation, that "commentaries are commonly more subject to cavill, than the Text; and therefore need other Commentaries; and so there will be no end of such Interpretation" (L 26.25, 193). I will return to Hobbes's understanding of historical writing in chapter 5, and to legal interpretation in chapter 6. Here, I want to briefly propose a Machiavellian background to Hobbes's objections to prudence.

In Machiavelli, the problem of the mediation of historical sources is a specific instance of a more general point, that political power itself is substantially a matter of signs and appearances. Thus, he argues in *The Prince*:

> To those who see and hear him [the ruler], he should seem to be exceptionally merciful, trustworthy, upright, humane and devout. And it is most necessary of all to seem devout. In these matters, most men judge more by their eyes than by their hands. For everyone is capable of seeing you, but few can touch you. Everyone can see what you appear to be, whereas few have direct experience of what you really are; and those few will not

> dare to challenge the popular view, sustained as it is by the majesty of the ruler's position. With regard to all human actions, and especially those of rulers, who cannot be called to account, men pay attention to the outcome. If a ruler, then, contrives to conquer, and to preserve the state, the means will always be judged to be honourable and be praised by everyone. For the common people are impressed by appearances and results. (P XVIII, 63)

Against those who think a ruler need actually *be* devout, Machiavelli's entire passage is about how the ruler should *appear*.[47] Indeed, the entire thrust of the discussion, if not the whole book, is to demonstrate to the ruler the overwhelming need to produce an appearance that keeps his subjects loyal and obedient: "above all, a ruler must contrive to achieve through all his actions the reputation of being a great man of outstanding intelligence" (P XXI, 77).

The point should be interpreted with caution. Machiavelli does not simply reverse some sort of Platonic preference for reality over appearances. The point is rather that he is a thoroughgoing materialist about power, and the symbolic environment is very much a part of the field of forces that a prince must successfully navigate in order to be successful. Appearances generate affects in the same way that armies do, and so their management is as important to the state as armies. Indeed, since the affect generated by an army is generated by its appearance, the appearance/reality distinction becomes increasingly irrelevant. Hence, the ruler "should avoid anything that will make him either hated or despised. If he does this he will have done what he should, and none of his other censurable faults will involve him in any danger" (P XIX, 63). It is not that the prince should avoid doing something bad; it is that he should avoid doing something which makes him look bad.[48] From the point of view of subjects, this produces an epistemic problem: how does one know whether one's ruler is good or not? By Hobbes, the question had been sufficiently generalized that he can write in the preface to *De Cive* that "we cannot tell the good and the bad apart." He then draws the result for practical reason, which has to act in the face of such uncertainty. "Even if there were fewer evil men than good men," he writes, "good, decent people would still be saddled with the constant need to watch, distrust, anticipate, and get the better of others, and to protect themselves by all possible means."[49] Machiavelli's caesura between appearance and reality thus paves the way for one of the constitutive elements of the Hobbesian state of nature, a universal mutual distrust.

Machiavelli's emphasis on appearances and phenomenality leads to a number of changes and even reversals from Aristotle on the question of practical reason. From a Hobbesian point of view, these will prove to be fatal to an effort to ground such reason in prudence. First, the endpoints.[50] According to Aristotle, prudence is a "disposition with true reason and ability for actions concerning what is good or bad for man" (NE 1140b5–6), and he immediately cites Pericles "and others like him" as examples of prudent individuals. A few paragraphs later, Aristotle establishes the framework for thinking about prudence in a political sense, noting that "prudence is not limited to what is universal but [one] must know also the particulars; for it is practical, and action is concerned with particulars" (NE 1141b15–16). He again emphasizes:

> It is evident, then, that prudence is not scientific knowledge; for it is concerned with the ultimate particular, as we said, and such is the object of action. It is thus opposed to intuition; for intuition is of definitions, for which there is no reasoning, while prudence is of the ultimate particular, which is an object not of science but of sensation, not the sensation of proper sensibles, but like that by which we sense that the ultimate particular in mathematics is a triangle. (NE 1142a24–8).

Here, then, Aristotle gives further warrant for his assertion at the beginning of the *Nicomachean Ethics* that politics does not, on account of the immense variability of human affairs, admit of great precision. In the context of historical texts and Machiavelli, the question can be sharpened: is this variability in human affairs so extreme that it swamps any effort to apply prudence to it whatsoever? The risk is that the heterogeneity of events, combined with the opacity of their recounting, swamps any effort to derive from them a law or regularity adequate to guide action. Hobbes concludes that it is better to start with the principle and not the events.

In attempting to account for the heterogeneity of events and limitations of their recording, Machiavelli offers an account of *virtù* that significantly departs from Aristotelian *phronesis* in several ways.

(1) For Machiavelli, in reading histories, one looks for people who authored imitable *acts*, rather than those who in general may be regarded as virtuous. For example, Machiavelli is able both to praise and to censure several rulers at different occasions in their careers, and he reads history according to whether a given person responded well to a given situation at the time it occurred. Hence, his reception of prudence involves rejecting the idea that "virtue . . . is a habit, disposed toward action by deliberate choice" (NE 1107a1); the Machiavellian virtue is in the act, not in the disposition.[51]

(2) The focus on acts alters Machiavelli's understanding of the Aristotelian mean. This point is perhaps most strikingly put in the *Discourses*, in a chapter explaining that a new prince should "make everything in that [newly acquired] state anew," where Machiavelli explicitly says that it is better to be an "bad" person than to search for the "middle" of an Aristotelian virtue:

> These methods are very cruel, and enemies to all government not merely Christian but human, and any man ought to avoid them and prefer to live a private life rather than to be a king who brings such ruin on men. Notwithstanding, a ruler who does not wish to take that first good way of lawful government, if he wishes to maintain himself, must enter upon this evil one. But men take certain middle ways that are very injurious; indeed, they are unable to be altogether good or altogether bad. (*Discourses* I.26)

Of interest here is not what the content of "good" and "bad" might be but the rejection of the mean. What Machiavelli emphatically rejects is the disposition "at the mean relative to us . . . between two vices, one by excess and the other by deficiency" (NE 1107a2). Machiavelli retains *phronesis*, then, but inverts the Aristotelian relationship between mean and excess/deficiency. That is, for Machiavelli, it is better to be at either "extreme" than at the "mean." One might put the principle thus: the average do not stand out. Those who do not stand out cannot rule because ruling depends on appearance. One should therefore decide to do something, *anything*, over indecision or indefinite action.[52]

(3) That prudence is a matter of action and not disposition necessarily generates a focus on the contexts in which a ruler acts.[53] I will indicate two aspects of this focus. First, the world will be presented as a series of threats to the ruler, which have either to be conquered or to be avoided. Success in this endeavor is highly specific to the threat, and so where an Aristotle can emphasize constancy of virtuous character, Machiavelli instead emphasizes adaptability as itself a virtue. If power is a matter of appearance, then one must add that appearance is always in terms of something and against a particular background. That is, things only appear within specific historical contexts, and an appearance of majesty can only be maintained with reference to the background against which it appears. One therefore both survives and becomes an example worthy of imitation by reading correctly one's own situation and responding appropriately to it. The entire focus remains with the ruler, and Machiavelli carries the point to its logical end, noting that "if it were possible to change one's character to suit the times

and circumstances, one would always be successful" (P XXV, 86).[54] Whatever the status of *fortuna* in Machiavelli's texts, if it conflicts with one's nature (singularly or collectively—one reason Machiavelli praises republics is their adaptability), and one is unable to adapt, fortune will gain the upper hand.[55]

If the ruler's final task is to adapt himself to uncontrollable circumstances, the median task is to correctly determine and regulate which circumstances are controllable. Machiavelli's ideal prince controls his own destiny to the maximum degree possible, but this control is always against a historical context which may be ultimately unchangeable. Hence:

> A man should never risk falling because he thinks it likely that he will be rescued. This may not happen, but even if it does it will not make you secure; such a defense is weak and cowardly, because it is outside your control. (P XXIV, 84)

Only those defenses that are under one's control and based on one's own *virtù* are effective, certain, and lasting. The focus on maximizing that which "depends on you" is recurrent. A few pages earlier, for example, he remarks that "it should be realized that all courses of action involve risks; for it is in the nature of things that when one tries to avoid one danger another is always encountered." This being the case, the ruler's task is again hermeneutic: "prudence consists in knowing how to assess the dangers, and to choose the least bad course of action as being the right one to follow" (P XXI, 79). A few pages prior to that, he says that "a wise ruler should rely on what is under his control, not on what is under the control of others" (P XVII, 61). What might otherwise seem a difficult example, then, becomes quite clear when viewed in this context: "I praise anyone who builds fortresses and anyone who does not, and I criticize anyone who relies on fortresses" (P XX, 76; *Art of War* Book 7 [pp. 183ff.] contains detailed instructions on how to build them). The important thing is not whether you have fortresses or not—it is that you must seem strong enough that you do not seem to rely on them, as building and then relying on fortresses is a sign of weakness, not of strength.

A second aspect of Machiavelli's emphasis on context is that the wise ruler, and the theoretical text addressed to a ruler who wishes to be wise, will make frequent use of historical examples. He writes:

> Nobody should be surprised if, in discussing completely new principalities, both as regards the ruler and the type of government, I shall cite

> remarkable men as examples. For men almost always follow in the footsteps of others, imitation being a leading principle of human behaviour. (P VI, 19)

Hence, "a ruler should read historical works, especially for the light they shed on the actions of eminent men" (P XIV, 53). Successful imitation in turn makes a ruler himself worthy of historical recounting, and reading historical works, Machiavelli immediately follows, is "above all, to imitate some eminent man, who himself set out to imitate some predecessor of his who was considered worthy of praise and glory" (P XIV, 53).[56]

That said, the nature of Machiavelli's move to example seems itself to undermine the historical knowledge on which it depends. The problem can perhaps be characterized as analogous to what will become the scholastic *regressus* problem.[57] Given that knowledge is of causes, and given that analysis—in this context, starting with the examples—only names the possible causes of a given effect, by what technique does one narrow the plenitude of possible causes of an effect to the correct cause? In the case presented here, the analogue is: given that activity is toward an end, how does one know the end structuring the activity of a historical example well enough to imitate it? We know *that* something happened, but we do not know it *propter quid*, according to its causes. Machiavelli imposes two further limitations on this imitation. First, he does not seem to be looking for a principle behind the examples, which implies that the examples may not even produce knowledge. Second, because human events, unlike nature, are infinitely variable, there is always going to be a gap between the present situation and the past. What one ends up with, then, is not so much a principle for action as an image whose structural components are opaque. Reverse engineering, here, is a necessarily imperfect activity.

Even more strikingly, that Machiavellian power is a matter of appearances has serious implications for the status of the ruler himself. If one follows the point about phenomenality to its logical conclusion, one arrives at the thought that one effect of Machiavelli's argument is that the person of the prince is himself contingent. This contingency operates at the most fundamental level: the problem is not one of the contingency of having this prince or that prince, which could in principle be resolved by historical research and the adequation of hereditary claims. The problem is that, with power unmoored from its ontological foundation in the piety of the ruler, it is immaterial whether anyone occupies the actual space behind the spectacle of power. If only the force relations matter, then it no longer matters whether there is an agent to initiate them, and it is contingent that any actual entity occupy the princely function. The materiality of appearance undermines the need to treat it as epiphenomenal.[58] The power of Machiavelli's prince is thus *arbitrary* in nearly every way. Against this backdrop, one aspect of Hobbes's thought becomes immediately clear: he tries to analyze the force relationships and then recuperate them into the persona of the sovereign, which is explicitly declared to be fictive and immanent to the commonwealth. At the same time, because he also radicalizes the thought that our perception of events is just that—perception—and therefore not amenable to easy corroboration with their actuality, he has to analyze those forces in a way independent of history and historical example.

I will return in the concluding chapter to the implications of Hobbes's incorporation of the contingency of the prince. For now, I want to indicate some of the contours of the Hobbesian response to Machiavellianism, by detailing in outline one aspect of this incorporation: the reduction of the sense in which power is understood as spectacle. Both general and specific traces of Machiavelli can be found in Hobbes.[59] The most obvious is in *Leviathan*'s discussion of justice. There, Hobbes notes that "Successful wickednesse hath obtained the name of Vertue: and some that in all other things have disallowed the violation of faith; yet have allowed it, when it is for the getting of a Kingdome" (L 15, 101).[60] I will develop a discussion of Hobbes's understanding of political thought and power in subsequent chapters. Here, I want to identify several indicia that the Hobbesian view is sharply different from Machiavelli's. For Machiavelli, as I have indicated, political thought is about a particular kind of prudence, which is executed in successful imitation of historical example, repetition of exemplary behavior, and adaptation to circumstances. In Hobbes, all of that is secondary to creating the circumstances in which one can act politically. The effect of this change is evident in their respective figurations of power, and the sense in which power is a matter of spectacle is dramatically diminished in Hobbes. One sees the effects of this diminution in at least three instances.

First, as an effect of the demotion of spectacle, historical examples are no longer dispositive for Hobbes. In the preface to his early Thucydides translation, Hobbes presents the orthodox Renaissance humanist reason for the study of history: "the principal and proper work of history being to instruct and enable men, by the knowledge of actions past, to bear themselves prudently in the present and providently towards the future" (EW VIII, vii).[61] By *Leviathan*, however, he is deeply suspicious of the study of other books. I have already noted the theoretical limitations Hobbes attaches to prudence; to that, one should add that this entails that books and examples are not to be studied without prior derivation of one's principles; as he puts it, "he that takes up conclusions on the trust of Authors, and doth not fetch them from the first Items in every Reckoning . . . loses his labour; and does not know any thing; but onely beleeveth" (L 5.4, 33). Study of history is even a cause of vainglory in the young; as Hobbes puts it, "The *vain-glory* which consisteth in the feigning or supposing of abilities in our selves, which we know are not, is most incident to young men, and nourished by the Histories, or Fictions of Gallant Persons" (L 6.41, 42–3). By the time he writes *Behemoth*, Hobbes is able to say unequivocally that "for the government of a commonwealth, neither wit, nor prudence, nor diligence, is enough, without infallible rules and the true science of equity and justice" (1990, p. 70).

Second, the image of sovereign power incorporates the power of its subjects in a way that moves a considerable distance toward a more modern, biopolitical configuration. I will discuss institutional manifestations of this movement in the chapter 6 and its implications for political philosophy more generally in the conclusion; here, I want to notice a visual image of it in *Leviathan*'s frontispiece. The printed frontispiece to *Leviathan* presents a deeply ambiguous image: a sword-bearing monarch rises above his kingdom, which lies before him as might a detailed model of it, including such features as hills, farms, and a walled town. The monarch's body armor is composed of an orderly crowd of faces, pointing inward and standing as if in docile submission to a disciplinary review parade. The frontispiece thus graphically represents a tension at the heart of Hobbes's definition of the commonwealth, that it is to be composed of the "real unity" of its subjects. Not only does the image not resolve, but also it emphasizes this tension in the duality of a sovereign rising over the city and the city below it, where the sovereign here is both on display and comprised the bodies of the populace. Much hinges on Hobbes's specific interpretation of the multitude and its transformation into a "people," but for now, one should note the incorporation of the people into the sovereignty. The original sketch for the piece betrays Hobbes's fascination with technique: in it, the people face outward, and the image bears a striking resemblance to one of Jean François Niceron's anamorphoses of Turkish sultans melding into the French monarch—images which could only be correctly seen from one perspective.[62] Hobbes's image thus invites his readers to adopt a correct theoretical perspective in order to resolve the faces into the monarch and see them as his body, but it leaves open the question of how the people could be elevated above their city as a first step toward this fusion. Why, in other words, do they compose the sovereignty, rather than being lorded over by it?

Finally, the exercise of the sovereign power in Hobbes is much less a matter of theatricality. To be sure, there are theatrical elements in the

Hobbesian schema. For example, he refers to the state of nature as a time when people lack a "visible Power to keep them in awe" (L 17.1, 117), though the remark needs to be interpreted with caution: the power to which he refers is that of the Leviathan, whose idealized image, as depicted on the frontispiece of the work is, as noted above, deeply ambiguous. When Hobbes discusses power in chapter 10 of *Leviathan*, he designates as natural power the "eminence of the Faculties of Body"; and says that reputational matters—including the "secret working of God, which men call Good Luck"—are instrumental to it (L 10.2, 62). The greatest of human powers, he adds, is obtained by combining the powers of various people into one person ("Naturall, or Civill") who has the use of their powers, "depending on his will." The commonwealth is the preeminent example of such augmentation, though Hobbes lists factions, alliances of factions, having friends, and having servants as examples of "strengths united" (L 10.3, 62). Reputation is a central aspect of obtaining this augmentation, and Hobbes offers what might be taken as a succinct précis of Machiavelli's point: having noted that "Reputation of Power, is Power, because it draweth with it the adhaerence of those that need protection" (L 10.3, 62), he adds that "what quality soever maketh a man beloved, or feared of many; or the reputation of such quality, is Power; because it is a means to have the assistance, and service of many" (L 10.7, 62).

Nonetheless, Hobbes moves the accent to questions of securing the cooperation of others, rather than their stunned submission. This shift of accent is evident in his treatment of capital punishment. Machiavelli frequently cites graphic and public executions as successful instances of sovereign power, with the usual intention of solidifying one's power base or impressing the multitude. One of the best examples is in chapter 7 of *The Prince*, when Caesar Borgia sends his "cruel and energetic" lieutenant Remirro de Orco to restore order in the Romagna. When Remirro grew in power and developed a reputation for cruelty, Borgia disposed of him:

> Later, the Duke considered that such great power was undesirable, because he was afraid it would incur hatred; and he set up a civil tribunal under a distinguished president, in the centre of the region, to which each city sent a lawyer. Because he recognized that the severe measures that had been taken had resulted in his becoming hated by some people, in order to dispel this ill-feeling and win everyone over to him, he wanted to show that if any cruel deeds had been committed they were attributable to the harshness of his governor, not to himself. And availing himself of an appropriate opportunity, one morning the Duke had Remirro

> placed in two pieces in the square of Cesena, with a block of wood and a blood-stained sword at his side. This terrible spectacle left the people both satisfied and amazed. (P VII, 26)[63]

In Dacres' seventeenth-century rendition, "he caus'd his head to be struck off one morning early in the Market place at *Cesena*, where he was left upon a Gibbet, with a bloody sword by his side" (1673, p. 545).[64] More generally, the prince is advised to execute as needed, without further reason or justification, without any grounds other than impressing people with his power. Such strategies have a purpose: "by punishing a very few he will really be more merciful than those who over-indulgently permit disorders to develop, with resultant killings and plunderings" (P XVII, 58). Borgia's genius in dispatching and then disposing of Remirro is that it allows him to appear merciful in his cruelty.

The situation in Hobbes is entirely different. It is true that the Hobbesian sovereign, by right, is able similarly to dispose of his subjects, without injustice, and I will return in chapter 6 to possible limits to the absolute power vested in a Hobbesian sovereign. Whatever limits might or might not exist, there is none of the emphasis on prophylactic executions. Also, the emphasis on spectacle disappears: Hobbesian violence is functional and restrained; his sovereign (as I will argue) is absolute but not arbitrary. Hence, in his late *Dialogue on the Common Laws*, the lawyer proposes a remedy for high treason entirely in keeping with Machiavellian spectacle:

> To be drawn upon a Hurdle from the Prison to the Gallows, and there to be hanged by the Neck, and laid upon the ground alive, and have his bowels taken out, and burnt, whilst he is yet living; to have his Head cut off, his Body to be divided into four parts, and his Head, and Quarters to be placed as the King shall assign. (1971, p. 143)

After questioning the justification according to reason of a number of such varying forms of execution, the philosopher finally concludes that "death being *ultimum supplicum*, is a satisfaction to the Law" (1971, p. 146). In sum, the Hobbesian account of punishment and executions can stand as an instance of the general point that I will pursue for the remainder of the book: Hobbes's understandings of power, reason, and politics present something quite innovative.

Chapter 3

More Geometrico

Laudamusque tuas, aeterne Geometer, artes.

Hobbes, OL V, 331

As I hope the previous discussion has shown (if only in outline), political philosophy is not a fully autonomous discipline. Hobbes's political philosophy is no exception, and the combination of Hobbes's aspirations to systematicity and his immersion in the context of the development of early modern philosophy more generally makes it especially important to attend to the metaphysical and epistemological underpinnings of his philosophy. In this chapter, I will examine one such underpinning: his geometry. Because Hobbesian mathematics is, for good reason, infrequently studied, it will first be necessary to set up some of the context in which this work appeared. I will then turn to analysis of the geometry, in particular of Hobbes's refusal to countenance symbolic algebra. This refusal, I argue, stems from a deeper methodological commitment on Hobbes's part to synthesis (rather than analysis) as a form of reasoning. This commitment, in turn, evidences a consistent concern on Hobbes's part with referentiality, and a rejection of any ontological import to referential statements.

3.1 "His Greatest Sanctuary"

Even a cursory glance at seventeenth-century sources suggests both that Hobbes viewed his own work as providing a geometric foundation for political philosophy and that he was read as attempting to provide such a foundation. The problem is that Hobbes loudly insisted on mathematical propositions that were, at far as any of his contemporaries could ascertain, incoherent. Hobbes's mathematician opponents used this incoherence to try to discredit his entire philosophy. The more Hobbes's propositions were refuted, the more audacious his next round of incoherent propositions would be. In 1666, for example, in the epistle to his latest effort, Hobbes declared that "of those who have held forth on the same things as I, either I alone am insane, or I alone am sane." In the face of such a self-immolation, readers who wanted to continue to study Hobbes's political and other writings had to detach them from the mathematics.[1] To appreciate why Hobbes's mathematical misadventures are nonetheless of interest, it is best to begin with some doxography.

The epistle dedicatory to *De Corpore* makes explicit Hobbes's endeavor to resituate the sciences according to both geometry and logic. After saying that, "to an attentive reader versed in the demonstrations of mathematicians . . . [the text is] clear and easy to understand," he proceeds to order the progress in the sciences, starting with the geometry of the ancients, and the "logic by which they were enabled to find out and demonstrate such excellent theorems" (OL I, 1; cf. DM I.1, 106). The epistle *ad Lectorem* repeats the move: Hobbes's reader is advised to "imitate the creation"; in the first part of the text, about logic, Hobbes will "set up the light of reason." The next two sections are concerned with definitions and geometry. It is only after this that Hobbes will consider human nature and, finally, society (OL I, nn). The formulation in *De Cive* is even more striking: "whatever in short distinguishes the modern world from the barbarity of the past, is almost wholly the gift of geometry," and "if the patterns of human action were known with the same certainty as the relations of magnitude in figures, ambition and greed . . . would be disarmed, and the human race would enjoy such secure peace . . . it seems unlikely that it would ever have to fight again" (OL II, 137). For Hobbes, at least, the order is clear: first logic, then geometry, *then* more traditional "political" topics. As the passage from *De Cive* makes clear, the stakes in establishing geometric foundations for political thought were none other than the possible success of that thought itself.

Hobbes's move to geometry was not unnoticed by his contemporaries, many of whom complained about it. To cite two examples: Eachard, in staging one of his anti-Hobbes dialogues, introduced the text by proclaiming that Hobbes "by a starch'd Mathematical method . . . had cheated some people into a vast opinion of himself" (1958).[2] As noted in section 2.1, the Earl of Clarendon similarly opined that Hobbes would "erect an engine of Government by the rules of Geometry, more infallible then Experience can ever find out" (1676, p. 117). Those who did not object to the fact *that* Hobbes used mathematics—usually mathematicians—almost universally objected to *how* he used it. In particular, Hobbes was involved in a long-running dispute with John Wallis, the Saville professor of mathematics at

Oxford. Both parties to that dispute saw their positions as having theological and political consequence, and so neither missed any opportunity for criticizing the other. As Hobbes wrote to his friend and translator Samuel Sorbière, "my quarrel with him is not like the quarrel between Gassendi and Morin or Descartes. I was dealing at the same time with all the ecclesiastics of England, on whose behalf Wallis wrote against me. Otherwise I would not consider him the least bit worthy of a reply" (HC 429). Wallis had similarly complained to Christian Huygens that "our Leviathan is furiously attacking and destroying our Universities . . . as though men could not understand religion if they did not understand Philosophy, nor Philosophy unless they knew Mathematics" and had found it "necessary that some mathematician should show him . . . how little he understands the Mathematics from which he takes his courage."[3]

In general, Hobbes's disputes with mathematicians grew steadily throughout his career, despite his having an early public reputation for geometric skills. Hence, in the 1640s, he was invited, along with Descartes, Roberval, and Cavalieri, to comment on a dispute between John Pell and Christian Severin Longomonatanus over squaring the circle (Probst, 1993, p. 275). By the mid-1650s, this reputation was entirely gone, and it had begun to fray well before that. Descartes had referred as early as 1641 to Hobbes's "most worthless ghost of a demonstration, in order to deceive the insufficiently attentive reader," and subsequently broken off correspondence with Hobbes partly because of Hobbes's refusal to concede algebraic proofs which seemed obvious to him.[4] By 1656, Mathematician Claude Mylon had written that neither he nor Huygens was able to find Hobbes's "thoughts about the dimension of the circle . . . comprehensible" (HC, 315). A year later, Mylon was begging Hobbes not to publish further on the subject "if you want to preserve your reputation," to "stop thinking about this subject . . . and apply yourself to more tractable matters" and to "spend your time more usefully than on this topic" (HC 479, 487, 490). Huygens wrote of his conviction that he was making the effort to refute Hobbes "utterly in vain—given that, in my opinion, he is incapable of being led thereby to admit his error,"—but nonetheless hoped he could persuade Hobbes to "abandon his extremely unsuccessful study of the whole of geometry" (HC 537, 538). In short, Hobbes's involvement in polemics about mathematics went well beyond his debates with Wallis. Hence, Wallis's judgment of 1662 was common, whether or not it was just:

> But now 'tis so unhappily found out, that *Geometry*, which he thought his greatest Sanctuary, hath most failed him. Nor is there any Tribe of men

> whatever, who are lesse satisfied with what he writes, than those who understand *Geometry*. . . . And doubtlesse, what ever else he is not, he is le[a]st of all found to be a Geometrician. (1662, pp. 6–7)[5]

The intensity and depth of these polemics suggest that Hobbes was insisting on something as a matter of principle. One therefore wonders: upon what was he insisting, and why did nobody else agree?

3.2 What is Hobbesian Mathematics?

This attention to geometry puts Hobbes in company with other early modern thinkers—Descartes and Spinoza come to mind—who at least nominally claim to proceed *more geometrico*, and whose work needs to be read through this claim. That said, the combination of Hobbes's subordination of geometry to logic and the virulent reactions of professional mathematicians suggests that what Hobbes means by geometric "method," and even by "mathematics," is not self-evident. Here, I will not claim that Hobbesian mathematics adds up to a "coherent program."[6] I will, however, reject the Whiggish sentiment that usually accompanies such an admission. There are clear foci in Hobbes's mathematical forays, and they evidence very specific metaphysical and epistemological commitments on Hobbes's part. It is the task of this chapter to uncover some of these commitments, and to show in a preliminary way why they matter for his political thought. Hobbes spent much of his career specifically opposing the efforts of the English algebraicists (of whom Wallis was the preeminent example), drawing a sharp distinction between geometry and algebra in the process. Hobbes's opposition to Wallis was total, but seems to reduce to two main theoretical points: (1) Hobbes rejected the move to symbolization in algebra; and (2) he consistently prioritized synthesis (reasoning from universal first principles to particulars) over analysis (reasoning upward from particulars to principles) as a method of obtaining knowledge, treating with hostile suspicion moves to establish parity between them.[7] As I will indicate, the issue is not so much academic skepticism as the proper definition of *scientia*.

The question of synthesis versus analysis is both logically prior to and dispositive for the answer to the one about geometry versus algebra, so I will start with synthesis and analysis. In chapter 20 of *De Corpore*, Hobbes defines the relevant terms in the context of geometry. He summarizes that "analysis is ratiocination from a supposition either constructed or made to the constructed or made efficient cause or many coefficient [causes]." Synthesis, on the other hand, is "ratiocination from the first cause of a construction through the middle [causes] and on to the made thing itself" (OL I, 254). The emphasis on construction or making should be evident at this point, and, in the context of geometry, Hobbes's interest is in genetic definitions.

A quick glance at the terms, however, suggests that there is a problem with parity. Analysis generates a series of causes which would be sufficient to produce something, and its accuracy can be verified by synthetically generating the desired figure. However, what analysis does not tell one is the way in which the supposited figure was actually generated. In cases such as geometric problems, where we provide the supposition ourselves, this is not a particular difficulty, since the end of the analysis is to demonstrate either "the construction of the problem as [something] possible or the detection of its impossibility" (OL I, 253). In other words, the point is to detect whether our supposition is possible or not. In the case where the effect is not just supposed but given, however, it is unclear what purpose analysis serves, since we already know the effect to be possible. At most, as Hobbes explains with regards to physics, we will know how it was possibly generated (DCo XXV.1).[8] The more difficult question is how it was actually generated, and analysis does not provide this information, since, as Hobbes also notes, "there are many ways through which the same thing can be generated or the same problem constructed." (OL I, 254). This plurality of ways leads to a diversity of geometric techniques, and, in the case of complicated problems, a rule for which method to use cannot be given in advance, so "success is dependent on cleverness, prior acquired knowledge, and even partly fortune." Of these factors, prior knowledge is the only one fully in the control of the geometer, and so "as far as someone is a good analyst after, he was a geometer before" (OL I, 255).[9]

Hobbes's comments about analysis and synthesis in geometry connect him to a larger scholastic problem about the status of the two as methods of obtaining knowledge more generally. Both, in principle, presented ways of knowing: analysis began with an experience in the world and reasoned back to its cause, and synthesis reasoned from causes to their effects. Almost everyone in the Aristotelian tradition agreed that synthesis—a demonstration *propter quid*—was legitimate. Consensus broke down over whether analysis—a demonstration *quia*—was in fact a true species of demonstration. Aquinas, for example, is willing to admit both forms; Avicenna rejects *demonstrationes quia*.[10] In schematic form, the problem with analysis dates to Aristotle and was glossed in the Coimbra edition of the *Posterior Analytics* as follows: "often we think we know perfectly, if we maintain a conclusion through some principle. However, this does not suffice, but it is required that the principles be proper to the thing which is demonstrated" (*In Lib. Aris. de Post. Res.* I.7, in Coimbra 1607, pp. 589–90). In other words, any number of principles or causes might account for the observed phenomenon, but only one of those can be the correct one, and so some sort of technique of reduction is necessary.[11]

In the seventeenth century, and despite his admiration for Galileo, Hobbes is notable for his unwillingness to entertain any strategy for resolving this *regressus* problem.[12] He alludes to this point in *De Homine*, where, as part of his explanation of why geometry and civil science are demonstrable but physics is not, he notes that "both of these methods of proof [cause to effect and effect to cause] are usually called demonstration; the former kind is, however, preferable . . . for it is better to know how we can best use present causes than to know the irrevocable past, whatever its nature" (DH 10.4). In one of his attacks on Wallis, Hobbes demotes *quia* from the realm of science altogether:

> I wish he [Wallis] would have defined this demonstration *quia*. For a demonstration *propter quid* is when someone shows according to what cause a subject has an affect. And although every demonstration is scientific, and knowing an affect is in the subject is by cognition of the cause which necessarily produces it, there is no demonstration other than *propter quid*. He rightly said, therefore, that which is called *quia* is not authoritative demonstration—that is, not a demonstration at all. (OL IV, 38)[13]

He then underlines that "in vain however do we seek the definition by demonstration *quia*, which is not a demonstration" (OL IV, 39). Hobbes follows this logic to its end: analysis proceeds by hypothesizing possible causes; even though the failure of the proposed effect to appear creates in the soul a psychological condition comparable to that induced by demonstration, "*scientifica non sit*" (OL IV, 39).

Algebra, then, exemplified the worst of all worlds: not only did it universalize the analytic art but also did it so without sufficient attention to its own method. Thus, in the *De Corpore* chapter, Hobbes quips that "symbolics, which is used in many ways today . . . is neither analysis nor synthesis," but a technique of "quickly gathering together the discoveries of geometry into commentaries." He then cautions that "although discourse between propositions greatly distant is easy with symbols, whether this discourse, when it is done without ideas of the things themselves, is of questionable use" (OL I, 257–8).[14] One of the characters in Hobbes's 1662 *Seven Philosophical Problems* issues the following triumphant pronouncement:

> I see you have wrested out of the hands of our antagonists this weapon of algebra, so as they can never make use of it again. Which I consider as a thing of much more consequence to the science of geometry, than either of the duplication of the cube, or the finding of two mean proportionals, or the quadrature of a circle, or all these problems put together. (EW VII, 68)

Given Hobbes's almost fanatical commitment to his own efforts at solving the problems he lists, the comment is remarkable. More important to Hobbes even than squaring the circle is the use of geometric demonstration in refuting the use of symbolic algebra in mathematics. That no one but Hobbes considered any of his demonstrations about the circle convincing should not allow us to be distracted from the problem they pose: on Hobbesian grounds, geometry is a science but algebra is not, and that seems to be because symbolic algebra's failure to attend to "the things themselves" makes it even less useful than other forms of analysis.

A second point of emphasis further muddies the water: even the claim that Hobbes followed a Euclidean model of geometry cannot be straightforwardly made.[15] In a 1666 text, Hobbes cites Clavius's introduction to Euclid and suggests that geometric method is not self-verifying:

> By means of which words he praises not the art of it, but its majesty. The certainty of all the sciences is equal, for otherwise they would not be sciences: for "to know" does not admit of more or less. Physics, ethics, politics, if they are to be well-demonstrated, are no less certain than the pronouncements of mathematics, just as mathematics is not more certain than the other sciences, unless that which it pronounces is correctly demonstrated. (OI IV, 390)

We will return to Hobbes's odd inclusion of "physics, ethics [and] politics." For now, note that the repeated emphasis on demonstrating "well" or "correctly" suggests the fundamental importance Hobbes assigned to logic and the correct specification of the synthetic art as a guarantor of certainty. This indicates that for Hobbes, the success of *De Cive* and his other political writings as political science hinges on refiguring politics as a demonstrable science on the model of classic geometry. It also implies that Hobbes will resist efforts to found any of these sciences on another basis.

Because Hobbes is concerned with methodological precision, he proceeds to comment on, and emend, Euclidean method. His discussion of the "point" is instructive. According to Euclid, says Hobbes, a point is "that of which there is no part [*cujus pars nulla est*]" (OL IV, 391). He then presents a view of symbolization which grounds a difficulty in Euclid's definition: "a sign is not the name of a quantum" (OL IV, 391). Euclid thus contains a hidden ambiguity:

> And these words . . . can be doubly understood: either as undivided; for a part is not understood, unless preceded by division; or as indivisible, that which is incapable of division by its nature. In the former sense, a point is rightly said to be a quantity; in the latter, it is not so, for all quantity is divisible. (OL IV, 391)

In this distinction, I think, lies the entirety of Hobbes's debate with the algebraists. For now, note the precision of Hobbes's solution to the ambiguity: "a point is indeed divisible, but no part of it is to be considered in a demonstration" (OL IV, 392).

Hobbes's attention to the act through which a geometer constructs proofs, and the artificiality of those constructed proofs as somehow *representing* nature without, at the same time, mirroring it, both marks the modernity of Hobbes's thought and shows a remarkable consistency between the theoretical concerns of his geometry and his politics. As I noted, he quite explicitly says in *The Elements of Law* that "one [way of erecting a body politic] is by arbitrary institution of many men assembled together, which is like a creation out of nothing by human wit" (EL 20.1). The Hobbesian analogy between geometry and politics should begin to be evident because in both cases instituting an axiomatic system first makes it possible to have the science in question. Elements of that system—whether citizens or points—have meaning only as parts of the system. For this reason, it is vital to attend to the process through which the system is constructed: misdefine a point, or the conditions for politics, and the entire system risks being unsound.

The modernity of the process is marked by its reliance on an understanding of thought as constructive or *poietic.* Considered as objects of knowledge, the objects under (scientific) consideration do not preexist their constitution as parts of the system; instead, thinking "brings forth" its objects, and the active presence of thought in the objects fundamentally separates them from "nature."[16] In other words, because the mind represents but does not mirror nature, the inquirer plays an active role in knowledge acquisition, which in this sense always builds an apparatus through which nature can be known. Correct method, then, attends to this process of construction; insofar as thought occurs in language (which for Hobbes it always does), method must attend to symbolization, frequently with an eye to limiting its scope or indicating that its results are possibly (because internally consistent, etc.) but not necessarily (because they are representative and therefore the representation itself could be erroneous) correct.

In terms of early modern understandings of calculation and numeration, this hybrid is anomalous, though not unique. Although clearly motivated by a different metaphysics, Kepler, one of the luminaries cited by Hobbes at the opening of *De Corpore*, comments that "nothing is proved by symbols . . . unless by some reason it can be demonstrated that they are not merely symbolic but are descriptions of the ways in which the two things are connected and of the causes of this connection."[17] More mainstream seventeenth-century thinkers such as Wallis, innovators in mathematical developments, expressed fewer or even no such worries about the scope of symbolization. On the other hand, thinkers who straightforwardly adhered to a premodern understanding of science tended to downplay the role of construction in knowledge acquisition. In this sense, and as Hobbes's emendation of Euclid on the definition of "point" suggests, Hobbesian demonstration also differs radically from a purely Greek model: for Hobbes, but not for the Greeks, a demonstration creates the objects it demonstrates. In both instances—politics and geometry—the specific art in question is to be subordinate to the logic of this production.

In this respect, and despite its vocal opposition to Aristotle, Hobbes's thought is concerned with order and method in the same way Aristotle's was.[18] Since it was in opposition to Wallis that Hobbes's mathematical ideas were most frequently expressed, a review of Wallis's work is in order. In particular, given the Hobbesian concern with order and method, Wallis's *Institutio Logicae* will begin to make the differences between them clear. Although against what he takes to be the abuses of scholasticism—in particular, separated essences—Wallis will nonetheless purport to follow Aristotle: "not that I am an innovator in this; but I will return to the doctrine of Aristotle those things (either sciences or, as I had thought better, non-sciences) which they had dragged out of it" (1687, *Dedicatio*, nn). Wallis is also, to an extent, a nominalist: "there are man, cow, horse, sheep, etc. . . . which by common consent signify (among us) the animals which we are accustomed to indicate with these names" (1687, p. 2). This suggests both Wallis's proximity to and distance from Hobbes. In particular, although Hobbes shares the nominalism, he is deeply suspicious of the regulation of meaning *communi consensu*.

From the point of view of geometry, matters come to a head in the definition of "quantity." On Wallis's account, quantity will be predicated of an attribute (such as width), at the same time apparently also providing information about the object whose width is measured. The move is a necessary component of treating mathematics symbolically, insofar as symbolization requires that a sign refer to something other than its obvious referent. Still, Wallis provides no apparent apparatus through which this relation between the attribute and object could be regulated. It appears, on the contrary, to be somewhat *ad hoc.* He writes in a passage, the slippage of which is remarkable:

> *Quantity* (universally taken) is the notion (or concept of the mind) according to which we are accustomed to ask of a given thing by means of the comparative adverb "how." Thus, how long? How wide? How large? How heavy? How long a time? How great? How much? How many? Etc. And it is thus something about which we can properly inquire; everyone refers to it with this predication. (1687, p. 26)

Quantity becomes, in other words, and quite explicitly, the equalizer among the qualia: all of the qualia can be measured through quantity, which is to imply that quality can be successfully reduced to quantity, or, to put matters in a more seventeenth-century idiom, that all quantities are homogeneous. Hobbes will explicitly challenge this understanding of quantity.

In his first *Lesson* to Wallis, Hobbes says that "quantity" refers to "that which is signified by what we answer to him that asketh, *how much* any thing is" (EW VII, 192). He then cites Latin usage: the answer to "*quantum est* [how much]" is not "magnitude or quantity, but . . . *tantum, so much*" (EW VII, 192) As the *quantum . . . tantum* pairing suggests, quantity for Hobbes is a relational term, which is to say that quantity is always quantity of something and that the meaning of the quantity is always determined by the thing of which it is the quantity. On the one hand, this understanding suggests that for Hobbes, as for the Greeks, the question of calculation is ultimately a question of counting and thus is indissociable from the question of what is being counted.[19] A number is not a sufficient referent because the number as a mark can indiscriminately refer to many different things: "this putting off an unit sometimes for one line, sometimes for one square, must needs mar the reckoning," and "it can be no otherwise when you so apply arithmetic to geometry" (EW VII, 64). Hence, the English algebraists court absurdity, "for the same number is sometimes so many lines, sometimes so many planes, and sometimes so many solids," and "any arithmetical account used in geometry" is false, "unless the numbers be always so many lines, or always so many superfices, or always so many solids" (EW VII, 59–60).

On the other hand, that quantity is to be relational suggests that the Hobbesian attention will always be on *both* the discursive or conceptual space within which something is constructed and on the objects of that construction. In other words, Hobbesian attention will be focused not just on the appearance of something but on the rules under which it appears. In a slightly different context, he will even say that "*to fainesthai* [appearing] itself" is the "most admirable" of natural phenomena (DCo XXV.1). Hobbes will also explicitly caution against the reification of numbers which results from inattention to the nature of quantity as relational: "*Differe and the quantity by which they differ,*" he explains to Wallis, "are quite of another kind. *Differe . . . differing, exceeding,* is not quantity, but relation. But the quantity by which they differ is always a certain and determined quantity" (EW VII, 384). Hence, "it is necessary to the science of geometry to define what quantities are of one and the same kind, which they call *homogeneous.*" This is because "*Homogeneous* quantities are those which may be compared by . . . application of their measures to one another; so that solids and superfices are heterogeneous quantities, because there is no coincidence or application of those two dimensions" (EW VII, 198). In other words, numeration is radically incomplete without an explanation of how symbols refer, an explanation that must take into account that identical-looking symbols can refer to completely different things.

All of this underscores that, in seventeenth-century terms, a distinction of quantity is for Hobbes a distinction of reason. Because it is a distinction of reason, it is necessary to carefully specify the units through which the distinction of reason is to be made intelligible. I will pursue the second of these points in the following section ("The Ontology of Number"). Here, I want to underscore the way that Wallis' treatment of similar questions would appear inadequate from a Hobbesian point of view. The inadequacies are important because, as I hope to indicate, the differences between Wallis and Hobbes hinge substantially on the degree to which Hobbes radicalizes a set of nominalistic commitments that he otherwise shares with Wallis. Both thinkers, in other words, are *moderni,* in favor of the new sciences and the productive powers of new mathematics. In the following chapter, I will consider Hobbes's nominalism as it has to do with his understanding of signification; here, I will consider it as it affects his geometry.

A pair of appendices to Wallis' *Logic* attempt to reduce quantity to a distinction of reason or, what is the same, to declare that quantity and matter are different only by reason. The closing passage of the text argues that "I conclude, therefore, that matter and quantity are not really distinguished among themselves, as thing and thing, and probably not modally either; but only by reason [*ratione ratiocinata*], that is, that by an inadequate conception of the thing" (1687, p. 262). Wallis' terminology here is quite precise;

a review of its context should help to focus the differences between Wallis and Hobbes. The three basic kinds of distinction—real, modal, and by reason—were established by Suárez, who in turn was attempting to clarify ambiguities in the Scotist account of individuation. There were two problems. The first, and the one to which Scotus addressed himself, was how to establish a principle of individuation, through which one would know why two distinct things were really distinct. Distinction in this sense indicates a sense of ontological independence. The second problem concerned the degree to which distinctions perceived by reason had ontological counterparts. In general, two things which were really distinct could separately exist from one another whereas things which were distinct only by reason were actually the same thing, considered under different aspects. Hobbes's emendation of the Euclidean point from "that which has no parts" to that whose parts are not considered, as well as his bracketing of whatever differences might exist between individuals in the state of nature, are paradigm cases of the reduction of real distinctions to distinctions of reason. Importantly, in both cases, Hobbes does not deny that the real distinctions exist; he denies that they are relevant to the matter at hand. In this, he follows early modern precedent (in thinkers such as Descartes and Gassendi) in separating questions of epistemology from metaphysics.

The modal distinction attempted a middle point between a distinction of reason and a real distinction. As Descartes presented it:

> A *modal distinction* can be taken in two ways: firstly, as a distinction between a mode, properly so called, and the substance of which it is a mode; and secondly, as a distinction between two modes of the same substance. The first kind of modal distinction can be recognized from the fact that we can clearly perceive a substance apart from the mode which we say differs from it, whereas we cannot, conversely, understand the mode apart from the substance. . . . The second kind of modal distinction is recognized from the fact that we are able to arrive at knowledge of one mode apart from another, and *vice versa*, whereas we cannot know either mode apart from the substance in which they both inhere. (*Principles* I.61)

Descartes lifts Suárez's discussion almost verbatim; Suárez in turn develops the modal distinction to deal with ambiguities in Scotus' formal distinction. As he says, "Scotus does not explain with sufficient clarity whether this distinction, which he himself calls formal, is actual in the real order or merely fundamental or virtual" (1947, 1.13; cf. 1947, 2.5).[20] Scotus' problem was basic: Aristotelian principles of rational intelligibility reached only as far as species difference. However, it was clear that difference was also between individuals. That is, Aristotelian genus and species categories seem only to allow rational access to tokens of a type but not to actual individuals. Since one nonetheless knows that two members of the same species are nonidentical, one is entitled to ask for the grounds of their nonidentity. However, it is not clear at this point how to proceed, since species presents the lowest level of rational intelligibility. Scotus cites Avicenna, who is said to have put it this way: "horsehood is only horsehood—it is of itself neither one nor plural, neither universal nor particular."[21] For example, one knows why Socrates and a mule are different and why a person and a mule are different, but those reasons cannot explain the difference between Socrates and Plato. By the seventeenth century, the problem seemed particularly intractable. "It is a wonder how much philosophers and theologians disagree among themselves about this matter," declares Eustachius a Sancto Paolo in introducing his very brief survey of the *doxa*.[22] Two basic strategies had emerged: a Scotist and a nominalist.[23] Both Wallis and Hobbes are nominalists in one sense, but they depart in another.

The nominalist response to individuation was to deny that there was a problem: only singular things exist, as the dictum went, which meant that whatever categories and descriptive distinctions one developed between them were entirely of reason. For Ockham and the nominalists after him, categorical propositions were therefore equivalent to concatenated instances of individual predication. In other words, the proposition that "nature is x" means "this nature is x, that nature is x. . . ."[24] In this regard, Ockham is content with saying that the members of a species are "diverse" but not "different" in the sense of genus/species (*Op. Th.* II, 212–14). For Ockham, the question of the numeric unity of individuation—of why something is what it is, and not something else—is a question of predication and denomination; the metaphysical question is "immediately" resolved, which is to say that it never existed in the first place. He writes:

> I say that there is no unity existing by nature in this stone which is not equally primarily in that stone. For I distinguish two kinds of unity. In one way, unity is said according to that which precisely denominates something as one and not plural, not one in comparison to something distinct from it in reality; in this way I say that every real unity is a numeric unity. In the other way unity is said according to that which denominates several things, either one in comparison to another thing distinct in reality, and in that way specific unity denominates Socrates and Plato themselves, and generic unity denominates this man and that donkey, not something

which in some way is distinct from this individual, but immediately denominates the individual itself. (*Op. Th.* II, 202–3)

In sum, not difference, but unity beyond the facticially existing singulars is the problem. Hence, "everything outside the mind is of itself 'this'; one should not ask in what way this individuation is caused . . . but it would be better to ask the cause by which it is possible that there is something common and universal" (*Op. Th.* II, 197). Ockham therefore has no place or need for the formal distinction or its modal successor: "it is impossible for created things to differ formally unless they are really distinguished; therefore, if nature is in some way distinguished from the contracting difference, it is necessary that they are distinguished as thing and thing, or as entity of reason and entity of reason, or as real entity and entity of reason" (*Ord* I d. 2 q. 6, *Op. Th.* II, 173).

Against this background, Wallis appears as an Ockhamite nominalist, whereas Hobbes appears as an Ockhamite nominalist who refused, on grounds based on his psychology (see the following chapter) to concede that matter might actually be as we perceive it. In his *Logic,* Wallis first reduces quantity to modal difference. He begins by citing Aquinas: "by quantum of a thing I understand matter, which is the adequate subject of quantity, to which it first inheres" (1687, p. 256). Having located quantity primarily in both number and matter, the proof is straightforward. He cites Suárez on the propositions that "quantity does not differ from the number of a substance, except modally"; and "that of which the modes do not really differ do not themselves really differ; and the mode of substance and the mode of quantity do not really differ; therefore they do not themselves [really differ]." He then concludes that "substance and quantity are partial subjects of their mode by number" (1687, pp. 257, 259).

Wallis does not content himself with this, and his "second assertion" is that "it is not probable that substance and quantity modally differ, but only by reason" (1687, p. 260). Although this is a dialectical thesis in Aristotle's sense, its verification would have striking consequences. The effect would be that continuous quantity turns out to be discrete, with the result being a sort of atomism that guarantees the intelligibility of the universe through number: Wallis's priority of arithmetic over geometry naturalizes itself, in transforming from a question of method into a question of metaphysics. The nominalist denial of separated essences, in other words, can be said in two ways. For Hobbes, it is to point to the fundamentally political nature of signification and the seditious confusions which result when signification is confused with reality.[25] For Wallis, it is also to point to the confusions which result when signification is confused with reality but only to then suggest that reality can itself be discussed in the politically neutral language of quantity, since quantity inheres equally in all reality. From a Hobbesian point of view, the problem emerges in an explicit confusion in Wallis's original citation of Aquinas, in which matter and quantum were taken to be equivalent terms. This equivocation precisely assumes what is to be proved, namely, that matter is a discrete, not a continuous quantity; that is, that principles of rational intelligibility apply straightforwardly to nature itself.

Of the confusion of method and metaphysics—or, rather, of essentializing moves in metaphysics—Hobbes has the following to say:

> There is a certain *Philosophia prima*, on which all other Philosophy ought to depend; and consisteth principally, in right limiting of the significations of such Appellations, or Names, as are of all others the most Universall: Which Limitations serve to avoid ambiguity, and aequivocation in Reasoning, and are commonly called Definitions; such as are the Definitions of Body [etc.] . . . The Explication . . . of which, and the Terms, is commonly in the Schools called *Metaphysiques*; as being a part of the Philosophy of Aristotle, which hath that for title; but it is in another sense; for there is signifieth as much, as *Books written, or placed after his naturall Philosophy*: but the Schools take them for *Books of supernaturall Philosophy*. (L 46.14, 463)[26]

The point which Hobbes is trying to make here is the same one that he made against Euclid: a point is something which is taken as such for the purposes of constructing a demonstration, but that usage of a point as axiomatic does not then imply that axioms are themselves part of nature. Hence, just as the original contract is constitutive of (and therefore prior to) the differentiation of the citizens in a commonwealth, so too is the original declaration of unity a logically prior act to the differentiation of the numbers in counting.

3.3 The Political Ontology of Numbers

Hobbes's commitment to geometry is thus tied to his understanding of logic and the correct method of demonstration; at the intersection of these two is a question about what one is doing when one invokes a number. In particular, Hobbes wants to know what the number is "of." In this section, I will pursue some of the metaphysical implications of this concern as they emerge in Hobbes's debate with Wallis. Recall Hobbes's commitments: the universe is composed of matter, which is continuous quantity. The correct science for the study of physics, then, is geometry, since it is the science of continuous quantity. Any numeration of this quantity—and this is where Hobbes parts company with both Wallis and Euclid—involves an act of construction on the part of the knower; this act should not be confused with a statement about the nature of what one is talking about. Since conscious construction is central to the endeavor, quantity becomes fundamentally a referential function, and any expression of quantity requires symbolization. Since numeration involves symbolization, and since symbolization is equivocal, definitions are important; otherwise, the results are meaningless as science.

In establishing, at least as a matter of probability, that quantity and matter cannot be distinguished ontologically, Wallis attempts to bypass all of these concerns. As various commentators have observed, results are more important than precision for the algebraists, and in fact many of their results did not receive rigorous demonstration until much, much later.[27] The effect of Wallis's move is a collapse of continuous and discrete quantity as a matter of metaphysics, which allows Wallis to substitute arithmetic for geometry: the universe becomes composed of discrete, not continuous, quantity. The attendant move to algebra means that everything hinges on symbolization. Accordingly, Wallis laboriously derives numeration and symbols in the opening sections of his *Mathesis Universalis*. Since he will both push the question of the priority of continuous and discrete quantity into the metaphysical background and establish the identity between a point as a unit of discrete quantity and the number "1," Wallis is in a position to derive a system of numeration which is complete and adequate. He secures the system of numeration against charges of equivocacy by establishing its pedigree in ancient texts—a move parallel to the "common consent" by which the meaning of words is secured. The ultimate sufficiency of the move is probably secured theologically: he derives the Latin numbers from the Greek and Hebrew (ch. 6), and then (ch. 7) derives the writing of numbers from the Hebrew alphabet.

That Biblical Hebrew anchors the system not once but twice suggests that Wallis is relying upon something like Adamite naming as an actual, anchoring event. The question of the Tower of Babel can be avoided by showing the correspondence of Greek and Latin numeration with the Biblical Hebrew. All of this underscores at least two points. First, Wallis joins the algebraists in stressing results over the precision of his system and is much more concerned with the application of his method than its strict justification. Insofar as Hobbes both insists on classical models of precision and distrusts anything which sounds like the application of theology where it does not belong, their collision was inevitable. Second, one should pause to underscore the radicality of Hobbes's nominalism, as it suggests just how far he will go in resisting anything which sounds like the confusion of thinking and nature, whether the topic is mathematics or politics. For example, in *Leviathan*, Hobbes explicitly cites the tower of Babel as evidence for the irreducibly political nature of language, thereby dismissing out of hand the possibility of any sort of accessible universal grammar or language (L 4.2, 25). I will return to these concerns.

Somewhat more surprisingly, Hobbes also says not only that numbers are words but also that numeration is the paradigm case of signification through words. Since he has already said that all language and thus all signification through words is political, this new claim means both that mathematical questions are political questions and that precision in one's system of signification cannot be settled by appeal to common usage. He writes:

> The use of words in registring our thoughts, is in nothing so evident as in Numbering. A naturall foole that could never learn by heart the order of numerall words, as *one*, *two*, and *three*, may observe every stroak of the Clock, and nod to it, or say one, one, one; but can never know what houre it strikes. . . . And he that can tell ten, if he recite them out of order, will lose himselfe, and not know when he has done: Much lesse will he be able to adde, and subtract, and performe all other operations of Arithmetique. So that without words, there is no possibility of reckoning of Numbers; much less of Magnitudes, of Swiftnesse, of Force, and other things, the reckonings whereof are necessary to the being, or well-being of mankind. (L 4.10, 27)[28]

The natural fool, in other words, is capable of individuation—"one, one, one"—but not numeration. And, most important, they are radically separate operations.

Wallis's ambivalent endorsement of irrational numbers is exemplary of the sort of imprecision to which Hobbes objects. Ratio, says Wallis, is distributed into rational and irrational, such that "irrational is that which is not able to be shown by true numbers; as, in a square, the ratio of the side to the diagonal" (1657, p. 251). The alleged superiority of arithmetic to geometry hinges on this, since one can easily work with such ratios in Euclidean geometry. Wallis admits the possibility of irrational numbers by reading his prioritization of arithmetic quantity and numeration into the Greeks:

> Indeed λογω (beyond all other signification) also signifies considered or computed; from which logistic is the same as that which I calculate or compute; and, that αλογον, which flees calculation, is not explicable by a true number; which is the reason it is called both αρρατον and ανεκσφονατον, inexplicible, ineffable, unable to be explained in detail: not that it is altogether unable to be explained in detail, but not able to be explained with true numbers. (1657, p. 251)

Following the passage above, Wallis shows how to calculate with irrational numbers, showing how they are greater or less, multiples, and so forth. He even uses them in the calculation of proportion. Posterity, of course, justified Wallis's faith in his results, but from a Hobbesian point of view, the move is the height of absurdity: Wallis is ready to admit into his system, and then use, a symbol whose referent he knows to be impossible to specify. Hobbes puts the sum of the complaint very clearly: "since the beginning of the world there has not been, nor ever shall be, so much absurdity written in geometry" as in Wallis's books, "the cause whereof I imagine to be this, that he mistook the study of *symbols* for the study of *geometry*" (EW VII, 187).

On Hobbes's account, then, symbolic algebra fails to individuate its objects correctly. Number is always number of something, which means that part of a mathematical proof is the correct specification of its object domain. A proof which does not begin by such correct specification, or which slides between object domains through the symbolic function of numbers, is either uncertain or sophistic or both. If these considerations help to explain why Hobbes and symbolic algebra were (so to speak) heterogeneous quantities, they do lead to the following peculiarity: a distinctly antimodern understanding of mathematics gives rise to the first distinctly modern understanding of politics. That is, in claiming to turn the study of politics into a science, on the model of other new sciences, Hobbes insists on a model of geometry that he (correctly) traces to the ancient Greeks, a model which was generally abandoned by his scientifically minded contemporaries.[29]

The preceding comments concerning Hobbes's emphasis on construction but opposition to symbolization suggest that the Hobbesian innovation in political science, the one which makes it "modern," is not the social contract theory *simpliciter*. The state of nature and social contract turn out to be moments which are part of a larger constellation of thoughts; the Hobbesian innovation—the one that allows such a constellation—is that which declares that politics is a *demonstrable* science, as in the long passage dividing the arts into demonstrable and indemonstrable. The conviction that politics is demonstrable then generates the necessity to account for its phenomena in a regular way. This in turn generates the need to consider the members of a *polis* as indivisible units, as numbers of the same qualitative type. Civil philosophy will thus have to spend its labors on making sure that the members of the *polis* are, in all relevant senses, of the same qualitative type. One can only calculate with homogeneous quantities—which means that political philosophy must begin by establishing a referential field whose quantities are homogeneous, in the same way that a proof in geometry begins by specifying what will count as a point. This will have to be a primary determination; otherwise, political science would make the same mistakes as symbolic algebra by treating different things as the same.

What all of the above should indicate is that the Hobbesian treatment of the state of nature—in particular, his insistence that people be treated as if equal—is a result of his more general methodological commitments. For Hobbes, the determination that objects are qualitatively the same (homogeneous, therefore able to be part of the same geometry and set of demonstrations) will found a science, rather than be determined by it. That is, although Hobbes, in positing the idea that those in the state of nature should be considered as equal, will indict Aristotle for saying that some people are slaves by nature, his ability to do so is consequent upon his having already negated the Aristotelian proposition that one aim at the amount of precision appropriate to one's subject matter, with politics admitting of less precision than the natural sciences (NE 1094b12). In a certain sense, Hobbes is arguing that civil philosophy should not be a natural science. In order to be scientific at all, it will have to establish its autonomy from nature. And Hobbes finds it imperative that civil philosophy become scientific, because scientific discourse is as insulated as possible from endless controversy—which in politics can erupt into devastating and brutal civil war. The greater the possibility of certainty, the more insulated the discourse will be from controversy; hence both the model of classical geometry and the effort to fight symbolic algebra. Algebra eroded what on Hobbesian grounds was the most valuable feature of classical geometry: its synthetic, demonstrative method.

Chapter 4

Nominalism Redux

What will we now say, if indeed ratiocination is nothing other than the coupling and concatenation of names or appellations by this word "is"? From which [it follows that] we gather by reason absolutely nothing about the nature of things, but about the names of them, especially whether or not we couple the names of things according to agreements (which we make by our will with regard to the signification of them).

Hobbes, 3rd Set of Objections to Descartes' Meditations[1]

In an early essay, Leibniz says that Hobbes's nominalism is of an extreme variety. "Not content to reduce universals to names, as do the nominalists," Leibniz suggests, Hobbes "says the truth of the matter itself consists in names, and, moreover, depends on human will: the truth depends on the definition of terms, but the definition of terms depends on human will" (1875–1890, IV, 158).[2] Leibniz is correct: Hobbes does not just say these things; he says them unequivocally and repeatedly. There are two separate propositions; first, the reduction of universals to names; and second, the reduction of names to will. Since Hobbes also says that will reduces to affect, the second proposition is in fact the reduction of names to affect. It is here that Hobbes's putative extremism lies. I will return in detail to his comments deflating universals and to the psychology that enables the reduction of names to affect. I want to begin by noting that Hobbes's nominalism frames his entire conception of philosophy, from his earliest works forward. In *De Motu*, we read that "right ratiocination, which is the work of philosophy, is nothing more than the correct combination of true syllogisms into propositions. However the truth of a proposition consists from the right joining of names . . . from which one concludes that true philosophy is not possible, which does not have its basis in an adequate nomenclature of things" (DM XIV.1, 201–2). In *De Motu*, we can also see clearly the connection, discussed in the previous chapter, that Hobbes draws between geometry and nominalism. He proposes that if we want to "demonstrate the truth of some universal dictum, this is done first by explaining the name of the definition, as to exclude equivocal [words], that is to define it, then from the definitions weaving the necessary consequences, as the mathematicians do" (DM I.2, 106). (Good) mathematicians, in other words, always attend to the correct specification of terms.

In *Elements of Law*, he adumbrates the stakes of this view of language: "the invention of names hath been necessary for the drawing of men out of ignorance . . . so also hath it on the other side precipitated men into error" (EL 5.13). This is because true and false "adhere to propositions and language" (*ibid.*), and most people speak from habit rather than reason. After the accretion of so many errors, "*ratio*, now, is but *oratio*" and:

> I may in a manner conclude, that it is impossible to rectify so many errors of one man, as must needs proceed from those causes, without beginning anew from the very first grounds of all our knowledge, sense; and, instead of books, reading over orderly one's own conceptions: in which meaning I take *nosce teipsum* for a precept worthy of the reputation it hath gotten. (EL 5.14)

Hobbes's citation of the "know thyself" dictum should not obscure the unorthodox gloss he applies to it: because knowledge starts from sense, and because words have to do with the remembrance of sense, self-examination will not discover Platonic forms or universals in the soul; all it will discover is more words. In short, central to Hobbes's entire philosophy is one of its more neglected aspects, his understanding of signification and the signifying process.[3] This process is at once psychological and epistemological.[4]

In this chapter, I will argue that, according to Hobbes, we have no intellectual faculty in which a prediscursive mental language could inhere; reasoning itself thereby reduces to the imagination and to the signification of material marks. In introducing the passage with which I opened the chapter, Hobbes draws a distinction between imagining, "that is, having an idea," and "conceiving in the mind, that is, using a process of reasoning to infer that something is, or exists" (AT VII, 178; PWD II, 125). Descartes expresses dismissive surprise and retorts that reasoning is about "the things that are signified by the names" and that he is "surprised that the opposite view should occur to anyone" (AT VII, 178; PWD II, 126). In a seventeenth-century context, Descartes is correct to be surprised, and it is the anomaly of Hobbes's position which will frame the discussion here. In the first section, I will read Hobbes with and against scholastic nominalism as it is developed in William of Ockham. In the next section, I will examine Hobbes's engagement with

Descartes. The final section will look at some of the consequences of the Hobbesian position for his political thought.

4.1 "Absolutely Nothing about the Nature of Things": Radicalizing Ockham

Initially, one can say that the nominalist point is to refute the notion that our concepts come from external, universal forms. There are two primary aspects to the Ockhamite critique, both of which will be echoed, though not repeated, in Hobbes.

(a) Ockham thinks that the real existence of universals would impinge on divine power. This is because they would limit God's ability to alter our experiences in the world. Hobbes, for his part, repeatedly asserts the importance of divine power, as is particularly evident in his debate on free will with Bramhall. There, he defines piety as requiring that "we think as highly of his [God's] power as we can"; he adds that even prayer "is not to move but to honor God" (1999, §15, pp. 27–8). Hobbes's position is as emphatic as it is radical. Citing *Job*—"who art thou, O man, that interrogatest God?"—he argues that "the power of God alone without other help is sufficient justification of any action he does. That which men . . . call by the name of justice . . . is not that by which God Almighty's actions are to be measured or called just" (1999, §12, p. 22). In the following paragraph, Hobbes declares to be incomprehensible a whole set of theological distinctions normally used to avoid the implications of this position, and rests on the thought that he is willing to change his mind in the (unlikely) event that such distinctions can be made comprehensible. This is certainly not Ockham's position, but the usage of divine power to avoid theological complications induced by apparent metaphysical strictures on divine action, strictures grounded the presence of a transcendental structure of meaning for terms like "just," has a direct antecedent in thinkers like Ockham, who worked in an environment overdetermined by the 1277 condemnations.[5]

(b) Ockham thinks that such universals are superfluous. "Ockham's razor" names (for us) the mechanism he developed in this critique: one can have a perfectly adequate account of human knowledge without resorting to the convenience of hypostatizing new metaphysical entities. The Hobbesian parallel is particularly clear in *Leviathan*'s critique of separated essences. For example, Hobbes remarks of words like "Free-will," "White*nesse*, Round*nesse*, Magni*tude*, Quali*ty*, Corruptibili*ty*" and the like that "when men write whole volumes of such stuffe, are they not Mad, or intend to make others so" (L 8.27, 59)? The scholastic roots of this position are evident in *De Motu*, as for example in Hobbes's discussion of accidents and being, where he directly addresses Porphyry. Of *esse* and *accidens*, "I do not understand 'accident' as happening by chance, or as enumerated by Porphyry among the predicables, but 'accident' is what is called predicamental, that is, what is contradistinguished from substance." In other words, accidents *per definition* have no separate existence, so "*being* is nothing more than the accident of a body, as the way it is to be determined and distinguished by the understanding. And so we call *to move, to rest, to be white*, and similar accidents of the body, and we say *to inhere* of bodies, because they are diverse ways by which we understand bodies" (DM 27.1, 313). Metaphysics has a sharply reduced agenda, and "consisteth principally, in right limiting of the significations of such Appellations, or Names, as are of all others the most Universall" (L 46.14, 463).[6]

Although Hobbes clearly picks up these and other aspects of the nominalist program, he nonetheless departs from the orthodox medieval version in a number of ways. Essentially, Hobbes takes and radicalizes the Ockhamite critique of universals and then adds to it his mechanistic psychology. As a result, the intellectual faculty succumbs to Ockham's razor. Hobbes is left with the position that words and language turn out both to be thoroughly affective, and both occur in the imagination. This view of language contributes heavily to his political thought, as the repeated denunciations of seditious speech suggest: seditious speech moves people to embrace nonsensical ideas. Even clear speech is dangerous for those who are not able to understand. In the debate with Bramhall, for example, he gestures to arguments about exposing the vulgar to philosophy: "if we consider the greatest part of mankind not as they should be but as they are . . . the dispute of this question will rather hurt than help their piety. And therefore if his Lordship had not desired this answer, I should not have written on it, nor do I write it but in hopes your Lordship will keep it private" (1999, §14, p. 27). More theoretically, and with fewer inhibitions, Hobbes opens *Leviathan* with the statement that "concerning the Thoughts of man . . . *Singly*, they are every one a *Representation* or *Apparence*, of some quality, or other Accident of a body without us; which is commonly called an *Object*" (L 1.1, 13). The parallel passage in the Latin edition offers two clues toward its interpretation. First, "object" is "*objectum*," which suggests that the topic under consideration is the so-called "objective reality" of ideas, that is, their presence in the soul. Second, the Latin indicates that these ideas are generated by the imagination, as the equivocal "*apparitio sive representatio*" underscores (OL III, 5). In this, Hobbes has moved both with and against Ockhamite nominalism. Like Ockham,

Hobbes does not rely upon universals or separated essences *extra animam* to generate these ideas. Unlike Ockham, Hobbes does not rely on "intuition" either. A review of Ockham will indicate the distance Hobbes has traveled.

Ockham's ontology admits only of "singular" things in the world. Predication is the product of the manipulation of our ideas of those singular things.[7] The primary theoretical task is thus logical, and consists in discovering how correctly to manipulate these concepts. Hence, according to Ockham, the encounter with a singular thing provides an "evident intuition [*notitia evidens*]" of that thing, *qua* singular thing. This intuition then becomes the object or term which is manipulated in logic. In place of a theory of universals, Ockham provides both a treatment of signification which explains the referential logic through which putatively universal terms operate, and a theory of "supposition" that discusses the ways in which our thoughts can stand for extramental objects. Ockham assumes that this intuition generally corresponds accurately with what is in the world. The mechanism by which such a reliable intuition arises is perhaps not clear, but it is clear that the general veridicality of intuition distances Ockham from questions about psychology and our perceptive apparatus and allows him to focus on the logic of signification. As the opening pages of *Leviathan* indicate, this separation of psychology and logic is impossible for Hobbes.

Hobbes similarly deflates universals, noting in *Leviathan* that there is "nothing in the world Universall but names, for the things named, are every one of them Individual and Singular" (L 4.6, 26), a proposition repeated verbatim in the Latin edition (OL III, 24). In these discussions, Hobbes moves through territory that would be familiar to readers of scholastic texts; in particular, his examples—Peter and John, man, and tree—are those typical of scholastic commentaries. His discussion in *De Corpore* also refers to the scholastic primary and secondary intentions and then denies existence to the latter: "it is manifest that genus, species, [and] definition are not names of something else beyond the word and name." Metaphysics errs in including them as such, for "they are only our thoughts on the nature of our significations" (DCo II.10). What is perhaps not clear from the initial passage in *Leviathan* is the degree to which Hobbes's thought is entirely on the signification and names, and not on the objects *extra animam.* He puts the point in striking terms, even in his early *De Motu*:

> A thing is properly said to be neither universal nor singular, which [terms] are only the appelations of names. As a man is one thing, Socrates is one thing, but "man" is a universal name. Socrates [is] a singular name. Universal ought rightly be defined as something which is said, or predicated,

> of several things; singular as that which is only of one. Predicated and said, however, are not of the thing, but of the appellation. (DM II.6)

To Mersenne (for Descartes), he writes that "one must know that although the name 'man' is a common name (one, in fact, of the five names that Porphyry expounds in his *Isagoge*), every man is either Peter or Socrates or some other individual" (HC 108). General terms like "man" or "nature" are themselves indefinite and without ontological status. Hence, "nothing is universal but names. Nature, therefore, if it is something universal, is not [anything]."[8] He adds elsewhere that "this universality of one name to many things, hath been the cause that men think that the things themselves are universal." Such thoughts are confused: "there is nothing universal but names" (EL 5.6). In sum, "philosophy teacheth us how to range our words; but Aristotle's ranging them in his predicaments doth not teach philosophy" (EW VII, 238).[9]

The emphasis on name, rather than referent, is also an effect of Hobbes's psychology, and, unlike Ockham, he is unable to separate psychology and logic. Hence, in place of Ockham's "intuition," Hobbes substitutes "representation" and "appearance," which means that the Hobbesian account ultimately depends on his understanding of phenomenality. The second chapter of *Leviathan* will therefore be on imagination, and in it, Hobbes critiques a caricatured amalgam of scholastic views of the imaginative faculty. The "Schooles" teach:

> Some saying, that Imaginations rise of themselves, and have no cause: Others that they rise most commonly from the Will; and that Good thoughts are blown (inspired) into a man, by God; and Evill thoughts by the Divell: or that Good thoughts are powred (infused) into a man, by God, and Evill ones by the Divell. Some say the Senses receive the Species of things, and deliver them to the Common-sense; and the Common Sense delivers them over to the Fancy, and the Fancy to the Memory, and the Memory to the Judgment, like handing of things from one to another, with many words making nothing understood. (L 2.9, 19)

One should first note that the argument is politicized from the start. In the paragraph prior to the one on school teachings, Hobbes had directly linked the nominalist critique to political obedience: it "ought to be the work of the Schooles" to dispel people's superstitious beliefs in such views; if "this superstitious fear of Spirits were taken away, and with it, Prognostiques from Dreams, false Prophecies, and many other things depending thereon, by which, crafty ambitious persons abuse the simple people, men would be much more fitted than they are for civill Obedience" (L 2.8, 19). The essentials of a nominalist position are all here, and it is clear that Hobbes will also be deploying the critique in the service of his political philosophy.

The replacement of intuition with imagination is both the axis on which Hobbes's argument turns and (as the preceding remarks on Ockham suggest) the one which distinguishes it from mainstream scholastic nominalism.[10] The various elements of Hobbes's position come together most sharply in *De Homine*'s chapter on "speech and knowledge." There, discussing the difference between animals and humans, he explicitly links imagination and intellect, and the whole complex to language use: "thus other animals also lack intellect. For intellect is in fact imagination, but which arises from the settled signification of words [*est enim intellectus imaginatio quidem, sed quae oritur ex verborum significatione constituta*]" (DH 10.1; OL II, 89).[11] As implied by *Leviathan*'s reference to sensible species and the *sensus communis*, Hobbes's target is scholastic developments of Aristotle's faculty psychology. These psychologies generally separate imagination and intellect as faculties. Imagination is the effect of sense-perception and hence is unstable. Intellect—which is operative in the Ockhamite "intuition"—involves the acknowledgement of and understanding of universals. Since universals do not change, intellect does not either, except insofar as one might add knowledge to it. The obvious question is how one gets from bodily imagination to intellect. Aristotle is unfortunately not clear on the point:

> Neither are these habits [i.e., principles of science and of art] present in the soul [from the start] in any determinate way, nor do they come into being from other more known habits. [They arise] from sensation, like a reversal in battle brought about when one man makes a stand, then another, then a third, till a principle is attained; and the soul is of such a nature as to be capable of being affected in this way. (*Posterior Analytics* 100a10–15)

The text outlines the general direction of a strategy: repeated sensations become memory; repeated memories become a single experience; and from either experiences or previously obtained universals arises a principle of an art or a science. These passages, taken from the end of the *Posterior Analytics*, are, as one modern commentator put it, among "the most obscure in all of Aristotle's writings" and have generated a wide variety of interpretations, both historical and contemporary (Milton, 1987, p. 52).

One "Platonic" solution is to posit the autonomous existence of universals which then occupy the intellect. For this position, it follows that, in some respect, the human intellect is separable from the body, the intellect participates in the realm of universals, and in so doing achieves its divine or natural purpose. The problem then becomes how one has knowledge of existing things, since one's cognition of them *qua* existent arises through sensation but knowledge (of essence) arises through intellect. Insofar as intellect (knowledge of essences) and imagination (images of sensible things) are separate, and to the extent that intellection is nonmaterial, sensibilia become almost irrelevant except as a prompt: it matters that *some* images of red things prompt my knowledge of redness, but it does not matter which ones. As I will indicate, Ockham addresses this problem directly: one has intellection of singular things. For Ockham as well as his Platonic interlocutors, however, intellect trumps imagination.

That sort of account also presupposes the existence of a separate intellectual faculty, even among those who deny extramental existence to universals, as can be seen by scholastic commentaries on the chapter. I will take Zabarella as exemplary. A universal, he says, is "that, which is one beyond the singular things, and represents that common nature which in all [the] singular things is one and the same"; he then underscores that it is not something *extra animam*, as the "Platonists" have it (*In . . . Posteriores Analyticos*, 1272B). Transition between the imaginative faculty (*fantasia*) and the intellectual faculty (*intellectus*) is *via* the intermediation of "intelligible species," as follows: "for when the *fantasia* imagines an image retained in the memory, it produces a species in the intellect, whereby just as sense is moved by the thing itself, *fantasia* is moved by sense, and *intellectus* by *fantasia*" (*In . . . Posteriores Analyticos*, 1269D–E). That Zabarella's main innovation over Aristotle is the introduction of another term mostly suggests the difficulty in accounting for the transition between faculties; nonetheless, it is this transition which he emphasizes. Hence, when Aristotle speaks of coming to "rest," Zabarella explains that the term, analogous to its use in physics, is to indicate the end of a movement from one faculty of the soul to another: "I think that this 'rest' signifies the end of a transition from one faculty of the soul to another" (*In . . . Posteriores Analyticos*, 1270E–F).[12]

All of this underscores that when Hobbes says that intellect *is* imagination, he is rejecting an entire epistemology. What distinguishes imagination and intellect is not the presence of different faculties in the soul; it is that intellect operates through the "settled signification of words." Of course, Ockham also rejects aspects of this problematic in that he no longer relies on universals *extra animam* as a ground of intelligibility. However, from this

Ockham derives the thought that singular things are intelligible in themselves: intuitive cognition is intellective. In other words, like accounts that rely on universals, Ockham's reliance on intuition also subordinates imagination to intellect.[13] Intellect will thus be said to "naturally" signify; as he writes, "a conceptual term is an intention or affect of the soul somehow naturally signifying or consignifying, capable of being part of a mental proposition, and able to supposit for it" (SL I, 1). Language is added after this, and Ockham emphasizes that "first the concept naturally signifies something and, secondarily, voice signifies the same thing" (SL I, 1, pp. 19–20). The primacy accorded to natural signification is also manifest in Ockham's treatment of affect, intention, and concept as equivocal terms. The effect is to separate a space for mental concepts, which then can serve as a check on the proliferation of meanings:

> A concept or an affect of the soul naturally signifies whatever it signifies; however, a spoken or written term signifies nothing unless by the institution of will. From which follows another difference, namely that a spoken or written term is able to change what it signifies at will, whereas a conceptual term does not change what it signifies through any such [act of] will. (SL I, 1, 46–52)[14]

When Hobbes uses terms like "mental discourse" and "natural" signification, he means something entirely different. The direction of the argument is clear as early as *Elements of Law*, where Hobbes writes that "a NAME or APPELLATION therefore is the voice of a man, arbitrarily imposed, for a mark to bring to his mind some conception concerning the thing on which it is imposed" (EL 5.2). The recurrence of "mental discourse" in the English *Leviathan* is perhaps ambiguous: "by *Consequence,* or TRAYNE of Thoughts, I understand that succession of one Thought to another, which is called (to distinguish it from Discourse in words) *Mentall Discourse*" (L 3.1, 20). The ambiguity disappears in the Latin edition, which explicitly reduces mental discourse to imagination. As he writes, "by the series of imagination I understand the succession of one cogitation to another; which, to distinguish it from discourse of words, I call *mental discourse*" (OL III, 14). In *De Corpore,* he suggests that language arises from the need to remember sense perceptions and stabilize cognitions from their natural "in flux and perishable [*fluxae et caducae*]" state. For this purpose, knowledge needs to acquire "some sensibilia as little monuments [*monimenta aliqua sensibilia*]." These "marks [*notae*]" are "sensible things added by our will, such that, by the sense of them, they are able to recall in the mind things similar to those thought, for the sake of which they are added" (DCo II.1).[15] A "natural sign" for Hobbes, then, arises from the habit of conjoining sensibilia to one another. In this way, natural signs reduce to conventional signs insofar as both are habituated. The difference is that a natural sign *qua* material signifier is similar in some way to the image signified, whereas conventional signs are the result of the imposition of will and carry no (or at most contingent) similarity to the images they signify. The relation between signifier and signified is thus arbitrary in the precise sense that it is instituted artificially and not by nature. As I will indicate, what emerges is thus a problem of how to reduce the impact of this arbitrariness by getting people to use the same system of signification.

Hobbes applies the point to accidents as well: accidents are what we perceive—indeed, apparition is "*alicujus qualitas vel accidentis in corpore externo*" (OL III, 5)—and on that basis, we infer the existence of objects in which such accidents inhere. None of this implies that the qualities in question are *real*: "there is in the object itself nothing more than the motion of matter, by means of which the object works on the sense organs in various ways" (OL III, 6).[16] Hence, when I imagine "white," I imagine a white thing, which serves as a mark for another white thing, based on the resemblance of the respective imaginings. There are of course indefinitely many properties in any object which I could pick out; that we focus on color is the function of habit and convention.[17] Accidents are singular and become common by convention. Hence, "*white* is therefore the name of a body subsisting *per se*, not of a color" (OL III, 528).[18] In consequence, "Aristotle errs, in that he did not distinguish between separate things and the separate considerations of a thing" (OL III, 531).[19]

Hobbes's account is thoroughly deflationary, and Ockham's intellective "*notitia*" becomes the graphic and material "*nota*." Since all knowledge is affective and bodily, no extramental universal could possibly be relevant to it, and since signification will be explained with reference to imagination, such universals will also be unnecessary. Furthermore, whereas Ockham will cryptically claim that the intuition arises "at once [*statim*]," thereby inviting (or at least not obviously precluding) accounts that rely on intelligible species as an explanation of how the signification in intellect is "natural," Hobbes provides a physicalistic account of the emergence of *notae* in the imagination.[20] This is the central issue: because, for Ockham, intellect is separate from imagination, he is able to separate a discussion of signification from one of perception. Having collapsed intellect into imagination, Hobbes has to speak of both signification and perception at once. Hence, for Ockham, words are instituted at will but checked by natural signification and mental discourse, whereas Hobbes says that signification is entirely a matter of will and of the addition of a mnemonic object to the imagination.[21] The mark is not a concept, as its materiality in the imagination suggests. In other words, thinking in this sense *is* linguistic; language is not something added later, "*understanding* being nothing else, but conception caused by Speech" (L 4.22, 30).

4.2 Hobbes contra Descartes

That Hobbes is targeting even the last vestiges of the intellect as a separate, nonimaginative, nonlinguistic ("intuitive") faculty is also evident in his response to Descartes' *Meditations.* From a Hobbesian point of view, Descartes appears as a traditional Ockhamite, unable to justify the invocation of a separate *intellectus.* To the *res cogitans,* Hobbes responds:

> Correct . . . But when he [Descartes] adds "that is, mind, soul, intellect, reason," doubt arises. For it does not seem to be correct argumentation to say: "I am thinking," therefore "I am thought" or "I am understanding," therefore "I am intellection." For in the same way we would be able to say: "I am perambulating," therefore "I am perambulation." Descartes assumes that being a thinking thing, and intellection, which is the act of thinking, are the same; or at least that being a thinking thing is the same as intellect, which is the capacity for thinking.[22]

In other words, Descartes is accused of reifying the act of intellection. Whether this charge is just is less important than the direction which Hobbes takes it. From the thought that we cannot conceive of an act without a subject, he suggests that "it seems to follow . . . that a thinking thing is something corporeal; for the subject of any act can be understood only in terms of something corporeal or in terms of matter" (PWD II, 122; AT VII, 172). The argument is against the proliferation of substances or essences: either we reduce substance to body and essence to signification or we end up naming as a substantial form the grammatical subject of every possible act. Against scholasticism, Hobbes was explicit: "once fallen into this Error of *Separated Essences,* they are thereby necessarily involved in many other absurdities that follow it" (L 46.19, 466), adding "one Inconvenience into another, without end, and without number" (L 46.23, 467).

The depth of the disagreement between Hobbes and Descartes emerges in Hobbes's fourth objection, where he suggests that thinking "is nothing other than the coupling and concatenation of names or appellations by this word 'is'" and that therefore "we gather by reason absolutely nothing about the nature of things, but about the names of them."[23] As he puts the point in *Leviathan*, reason "is nothing but *Reckoning* (that is, Adding and Substracting) of the Consequences of generall names agreed upon, for the *marking* and *signifying* of our thoughts" (L 5.2, 32). Descartes treats the opposite point of view as self-evident and responds curtly that "as for the linking together that occurs when we reason, this is not a linking of names but of the things that are signified by the names, and I am surprised that the opposite view should occur to anyone" (PWD II, 126; AT VII, 178).

The point to notice is that both Hobbes and Descartes treat the issue of whether language refers fundamentally to itself or to objects in the world as tied to the question of whether intellect can reduce to body and imagination. This point is confirmed in the fifth objection, when Hobbes denies that we can have an idea of God on the grounds that we can have no image of God. Descartes responds that "my critic wants the term 'idea' to be taken to refer simply to the images of material things which are depicted in the corporeal imagination." He adds that "I am taking the word 'idea' to refer to whatever is immediately perceived by the mind [*immediate a mente percipitur*]" (PWD II, 127; AT VII, 181). The Cartesian position should recall Ockham's reliance on the immediacy and self-evidence of intuition; Descartes defends himself as using "the standard philosophical term used to refer to the forms of perception belonging to the divine mind, even though we recognize that God does not possess any corporeal imagination" (*ibid.*). In other words, the separability of mind is undertaken in the service of theology as the guarantor of science, and Hobbes's effort is to account for thinking without reliance on the stabilizing apparatus of the divine mind. Hence, he will simultaneously have to develop a theology based on the unknowability of God[24] and an account of reason that can explain how a thinking consisting of nothing but phantasms can be sufficiently stable.

Hobbes repeatedly emphasizes the absence of natural meaning. In *De Corpore*, he notes that "it is to be supposed that names arose by human will" because "new words are daily born, old ones abolished, [and] diverse words are in use by diverse peoples." This fact precludes any natural signification or resemblance, and he asks rhetorically: "finally, who sees that there is any similarity between words and things, or is able to institute a comparison between them, or is able to conclude in his mind that the names themselves represented the very natures of the things themselves" (DCo II.4)?[25] As indicated above, and unlike medievals such as Ockham, for whom political philosophy began with the fall, the important Biblical referent for Hobbes is the tower of Babel. The passage above in *De Corpore* immediately refers to Babel, a reference repeated in *Leviathan*: "all this language gotten, and augmented by *Adam* and his posterity, was again lost at the tower of *Babel*, when by the hand of God, every man was stricken for his rebellion, with an oblivion of his former language" (L 4.2, 25). In other words, at the creation, something like Adamite naming perhaps guaranteed a linguistic realism, but after the tower of Babel, language is nominalistic.

How to settle the meaning of words, then? One way is through definitions, and this is why almost all of Hobbes's theoretical texts begin with an extensive catalog of definitions. As he explains, "in the right Definition of Names, lyes the first use of Speech; which is the Acquisition of Science: And in wrong, or no Definitions, lyes the first abuse; from which proceed all false and senslesse Tenents" (L 4.13, 28). He adds that there are four basic things which can be named, and thus four "generall" types of names: of matter or body; of accident or quality; of sensation; and when "we bring into account, consider, and give names, to *Names* themselves, and to *Speeches*: For, *generall, universall, speciall, aequivocall*, are names of Names" (L 4.18, 29–30). In sum,

> This is all the variety of Names *Positive*; which are put to mark somewhat which is in Nature, or may be feigned by the mind of man, as Bodies that are, or may be conceived to be; or of bodies, the Properties that are, or may be feigned to be, or Words and Speech. (L 4.18, 30)

After explaining that negatives "are notes to signifie that a word is not the name of the thing in question" (L 4.19, 30), Hobbes concludes that "all other Names, are but insignificant sounds; and those of two sorts." The first is "when they are new, and yet their meaning not explained by Definition," a common practice in the texts of "Schoolemen." The second is when names with contradictory significations are affirmed at once, as a "round quadrangle" or the like (L 4.20–1, 30). None of this implies any ontology or necessary structure of meaning. Rather, it implies just what Hobbes says it does, that "the manner how Speech serveth to the remembrance of the consequences of causes and effects, consisteth in the imposing of *Names*, and the *Connexion* of them" (L 4.5, 26). Cogitation occurs through signification, and signification is an act of imposition.

4.3 Scientia Civilis More Definitionis

Consideration of Hobbes's account of signification helps us to understand why his political philosophy takes the form it does and why it provoked such dismissive astonishment among his readers. Examples could be multiplied;

to Descartes, I want here to add two deliberately diverse instances. First, Eachard "did presume . . . to think his [Hobbes's] writings so fond and extravagant, as not to merit being opposed in good earnest" (1958). Among his targets is the Hobbesian account of language, and he complains that if one reads Hobbes's "Logick," one will "find a whole *Book* full of nothing but *new words*" (1958, p. 18). Second, in the Port Royal *Logic*, Arnauld and Nicole take considerable pains to deny Hobbes's response to Descartes. Ultimately, they reiterate Descartes' essential claim that reason is a "solid and practical judgment about the nature of things by considering ideas in the mind that people chose to mark by certain names." Their support of this conclusion also echoes Descartes: in addition to arguing on the basis of actual linguistic diversity, they claim that were there no ideas, the conventions on which agreement in language is based would be impossible as there would be nothing on which they could be established, "just as it is impossible to make blind people understand what the words 'red,' 'green,' and 'blue' mean by any convention because, lacking these ideas altogether, they cannot connect them to any sounds" (1996, I.1). From a Hobbesian point of view, this complaint misses the mark: since, for Hobbes, intellect reduces to imagination, the point is not that we will have no antecedent ideas; it is that these ideas are contingently similar products of the imagination. What we need is a replacement for the stabilizing universality of the intellectual faculty. Hobbes thus prioritizes definition in what Eachard correctly suspects involves the creation of numerous "new words."

Hobbes's account of political philosophy will thus substantially break both with any sense of a final causality derived from nature and with efforts to derive principles from custom or convention. Instead, he will begin with definitions, which will serve to anchor the branch of *scientia* concerned with the "consequences from the Accidents of *Politique* Bodies" (L 9 chart, 61). The chapter on discourse offers a possible source for Leibniz's remark and further sharpens the limitations on Hobbesian science. Even armed with correct definitions, knowledge of "matters of fact" is sharply limited:

> No Discourse whatsoever, can End in absolute knowledge of Fact, past, or to come. For, as for the knowledge of Fact, it is originally, Sense; and ever after, Memory. And for the knowledge of Consequence, which I have said before is called Science, it is not Absolute, but Conditionall. No man can know by Discourse, that is, or that, is, has been, or will be; which is to know absolutely: but onely, that if This be, That is; if This has been, That has been; if This shall be, That shall be: which is to know conditionally; and that not the consequence of one thing to another; but of one name of a thing, to another name of the same thing. (L 7.3, 47)

"Conclusion," then, will simply name the "End or last summe" of one's syllogisms, "and the thought of the mind by it signified, is that conditionall Knowledge, or knowledge of the consequence of words, which is commonly called SCIENCE" (L 7.4, 47–8), or as the Latin edition puts it even more succinctly: "and this *science* is indeed cognition of the consequences of one word to another" (OL III, 52).

"Conditional" thus has the sense of connected propositions; whether the knowledge achieves more than formal validity is a function of the definitions. Since we are the objects of our own political philosophy, the discipline carries both the chance to achieve certainty and the greatest risk of failure through badly constructed definitions. Hobbes's methodological point is thus that political philosophy needs to begin with careful consideration of its own terms, and the first things to avoid are therefore historically existing but speciously derived accounts of the ends of politics. For example, when he arrives at the chapter in *Leviathan* "on the difference of Manners," which contains the text's first presentation of "a perpetuall and restlesse desire of Power after power," Hobbes makes it clear that "the Felicity of this life, consisteth not in the repose of a mind satisfied. For there is no such *Finis ultimus,* (utmost ayme,) nor *Summum Bonum,* (greatest Good,) as is spoken of in the Books of the old Morall Philosophers" (L 11.1, 70).[26]

Hobbes will accordingly reformulate the *scientia* of living well. Living at all involves the use of language as an expression of human desire, and living well therefore involves using language well. Indeed, language is the *sine qua non* both of politics and living well. Speech is "the most noble and profitable invention" of distant antiquity, "without which, there had been amongst men, neither Common-wealth, nor Society, nor Contract, nor Peace, no more than amongst Lyons, Bears, and Wolves" (L 4.1, 24). Hence, where *De Cive* claims that people are both gods and wolves to each other, *Leviathan* makes it clear that language is the variable that determines which they will be. Since Hobbes thinks that metaphysical entities and separate faculties are nonsense, and since people express their desires through language, politics is about regulation of desire and not the imitation of universals. Indeed, to imagine people without imaginations and desires—and hence, language—is impossible. As he suggests, "nor can any man more live, whose Desires are at an end, than he, whose Senses and Imaginations are at a stand" (L 11.1, 70). Language, desire, and politics all indicate the same set of issues for Hobbes; the effort to separate them involves one in the metaphysical subtleties and theological niceties of fetishized concepts.

One may illustrate the far-reaching consequences of Hobbes's position with reference to one of its corollaries, namely that, for him, "good" and "evil" (and other moral words) have no nonpolitical referents. As noted above, this view emerges in the debate with Bramhall, where he rejects on theological grounds the possibility of measuring divine justice by the human concept. It is also evident in his complaint against the Calvinists having "privately" defined good and evil. In *De Cive*, after citing "certain Theologians in our own day" who believe that "tyrannicide is licit," he asks:

> If he holds power rightly, the divine question applies: *who told you that he was a Tyrant, unless you have eaten of the tree of which I told you not to eat?* For why do you call him a *Tyrant* whom God made a *King*, unless you, a private person, are claiming for yourself a knowledge of *good* and *evil*? (DC XII.3)

Since the Calvinist argument is, as the early Calvinist Theodore Beza (1970) put it, that a tyrant would be "entirely manifest [*toute manifeste*]" as a matter of knowledge or perhaps by an evident intuition, the point is not just to pit a public understanding of tyranny against a personal one. Rather, it is to indict private judgment in matters of public concern. For Hobbes, such declarations that one has "seen the light" are both dangerous and false. They are dangerous because they license any false prophet to try to overthrow the kingdom on specious religious grounds, and they are false because there is no possibility that the judgment "tyrannical" or "evil" refers to anything outside the public space of language. In other words, the word only has meaning in a public space, and since there is no extrapolitical standard of meaning against which to judge it, the right to define falls to the sovereign. This, of course, gives Hobbes the argument he needs: no intelligent sovereign would define himself as tyrannical, and so the judgment that a sovereign is tyrannical is always treasonous, and no appeal to an outside authority is possible.

Hobbes is absolutely clear that this is to be a general point. He writes in *Leviathan*:

> Whatsoever is the object of any mans Appetite or Desire; that is it, which he for his part calleth *Good*: And the object of his Hate, and Aversion, *Evill*; And of his Contempt, *Vile* and *Inconsiderable*. For these words of Good, Evill, and Contemptible, are ever used with relation to the person that useth them: There being nothing simply and absolutely so; nor any common Rule of Good and Evil, to be taken from the nature of the objects themselves; but from the Person of the man (where there is no Commonwealth;) or, (in a Commonwealth,) from the Person that representeth it;

> or from an Arbitrator or Judge, whom men disagreeing shall by consent set up, and make his sentence the Rule thereof. (L 6.7, 39)

Hence, "the notions of Right and Wrong, Justice and Injustice have . . . no place" in the state of nature; rather, "where there is no common Power, there is no Law: where no Law, no Injustice . . . Justice, and Injustice are none of the Faculties neither of the Body, nor Mind" (L 13.13, 90). This was merely an elaboration of the position he had taken as early as 1640, where, in *Elements of Law*, he declared that "the question, which is the better man, is determinable only in the estate of government and policy, though it be mistaken for a question of nature" (EL 17.1). In short, moral words are political words and are meaningless outside politics.[27]

Hobbes's understanding of signification is thus an integral part of his thought as a whole. From the empiricist dictum that all of our thoughts originate in sense impressions, he draws the further conclusion that this point applies to the thoughts we use to mark and communicate our other thoughts, that is, to language. In other words, there is no need to posit an intellectual faculty to understand human use of language. Like other early moderns, he further rejects the idea that the impressions in our own minds have any necessary relation of resemblance to objects in the world. The combination of these views marks a sharp break with scholastic nominalism. Evidence of this break is found at the level of Hobbes's constant attention to rhetoric and political speech. The absence of a stabilizing intellectual faculty—the reduction of intellection to imagination—makes it impossible for him to create a space for thought which is independent of the affects and the sociopolitical processes which move them. The primary task of political philosophy shifts from a derivation of terms from extrapolitical sources to the erection of a sovereign apparatus which itself performs this regulatory function.

Chapter 5

The State of Nature

Those who are concerned about the government should induce fears into the citizens and bring distant dangers near, thus making them not relax but, like night sentinels, keep watch and safeguard the government.

Aristotle, Politics *1308a28–31*

The philosophical project of Hobbes thus unfolds as essentially *poietic* or constructive, which implies that civil philosophy will be governed according to a set of rules which simultaneously generate and regulate its object domain. Such a civil philosophy rejects as insoluble the Machiavellian task of maintaining order against the natural world. For Hobbes, the principle that adaptation to the natural is impossible destroys the possibility of a political philosophy based on that adaptation. This, in turn, has the very real effect of radically destabilizing any subsequent conception of "sovereignty:" if the "sovereign" names the concept of order in politics, then it will not do to have sovereignty begin with deferral to external events. Actual sovereigns, of course, can and do constantly face such contingencies. But that is not the point. What is inadmissible is contingency at the theoretical level, that is, of positing a separate agent called "fortune" which disables the generation of political principles that can be applied without first locating proper historical examples.

The task that thus emerges for Hobbes is twofold. First, he has to establish the possibility of a conceptual space for politics as a *scientia*. Second, he has to guard that space against the intrusions of the world outside. This second step involves a series of elaborate immigration mechanisms designed to conjure away *doxa* and other demons that might threaten its possibility. Such mechanisms of course have implications for the prudential art of governance. However, they are not simply the application of a political theory developed in the first step. It would be more accurate to identify the second step with normative political theory and the first with a prior moment that establishes the terms and conditions which will be taken as axiomatic within it. These terms are derived from Hobbes's epistemological and metaphysical commitment to the constructive principles outlined in previous chapters. The greatest effect of Hobbes's reconfiguration of civil philosophy, as I have indicated, will be that it is no longer subalternate to an Aristotelian natural philosophy. In this chapter, I will consider the first step of this constructive process, Hobbes's posit that one should think in terms of, on the one hand, a state of nature and on the other, a political state characterized by being entirely ordered according to a sovereign power.

Since we are concerned with the Hobbesian description of the state of nature, it will be well to begin with a review of it here. The state of nature was a constant in Hobbes's major political writings, although it underwent a substantial evolution between *De Cive* and *Leviathan*.[1] *Elements of Law* reports that "the estate of hostility being such, as thereby nature is destroyed, and men kill one another" (EL 14.12). In the terms of *De Cive*, it is a "war which cannot be brought to an end by victory because of the equality of the contestants, [which] is by its nature perpetual," and "it must be regarded as a miracle if even the strongest survives to die of years and old age" (DC I.13). Both *Elements of Law* and *De Cive* adduce historical examples. In *Elements of Law*, we find the people "few and short-lived, and without the ornaments and comforts of life, which by peace and society are usually invented and procured" (EL 14.12). *De Cive* similarly offers examples of "nations, now civilized and flourishing, whose inhabitants then were few, savage, short-lived, poor and mean, and lacked all the comforts and amenities of life which *peace* and society afford" (DC I.13). By the famous passage in *Leviathan*, these examples have been elevated into an abstract but precise description:

> In such condition, there is no place for Industry; because the fruit thereof is uncertain: and consequently no Culture of the Earth; no Navigation, nor use of the commodities that may be imported by Sea; no commodious Building; no Instruments of moving, and removing such things as require much force; no Knowledge of the face of the Earth; no account of Time; no Arts; no Letters; no Society; and which is worst of all, continual feare, and the danger of violent death; And the life of man solitary, poore, nasty, brutish, and short. (L 13.9, 89)

As a matter of preliminary interpretation, one might venture that these are *negative* pictures and that Hobbes methodically deprives people in the state of nature of the accoutrements of society.[2] More important, they are deprived of all sources of *meaning*. Together, these suggest that signification and society are coextensive. Hence, the state of nature, figured by the dissolution of "people" into a "multitude," is a state of communal oblivion.

Since the state of nature as been the object of much competing discussion, it will perhaps be helpful to outline my argument and some of its central points. I argue that (a) Hobbes does not think the state of nature really existed. This conclusion is consistent with his insistence on equality as a methodological postulate and with his rejection of analysis in geometry. We are to consider people, in the formula of *De Cive*, "as if they had just emerged from the earth like mushrooms" (VIII.1); (b) He uses the state of nature, as one would expect from his geometry, to deduce what he takes to be the necessary conditions of a sovereignty. The state of nature serves the methodological function of leveling natural differences into a homogeneous matrix of intelligibility. The absolute disorder of the state of nature is thus written into it as an axiom, and I will emphasize the extent to which the totality of this disorder is paradoxically its criterion of intelligibility. (c) The state of nature would be a condition, among other things, of complete semiotic chaos. This result is expected from his radicalized nominalism and from the pervasive background influence of Machiavelli. (d) A primary job of the sovereignty is thus to restore communicative transparency; one avenue through which this occurs is in the various institutions that implement the sovereignty on a daily basis (I will explore this last point in greater detail in the next chapter).

5.1 "The mind of the reader is no less affected": Finding the State of Nature

Any oration, according to a loosely theorized but nonetheless broad consensus of the late Renaissance, needed to find "commonplace" arguments and maxims, to move the passions of its auditors and to establish the ethos of the speaker such that the audience would be willing to listen. Among the strategies for the latter are establishing the public importance of one's topic and the probity of the speaker's person. On this view, successful persuasion involves moving the audience's fickle passions to one's own point of view, as much by emotion as by reason.[3] It is certainly true that Hobbes deeply distrusted classical and Renaissance accounts of eloquence, but it is also the case that he was not above using those techniques to serve his purposes. In the English *Leviathan*, in particular, it is easy to notice a polemical purpose, in addition to a scientific one. As he puts it in the "Review and Conclusion," "wheresoever there is place for adorning and preferring of Errour, there is much more place for adorning and preferring of Truth" (L, R & C 4, 484).[4]

In this section, I want to analyze the various connotations suggested by Hobbes's images of the state of nature. Although I will argue that none of these images are actually intended to *be* the state of nature—that state cannot, *per definition* actually exist, and, in any case, one would never be able to know if it was in fact the state of nature—the images nonetheless function to arouse in *Leviathan*'s readers a fear of the natural condition and serve to provide a series of examples that serve as *problemata* or *loci cognitionem*.[5] Hobbes's rejection of any form of analytic method or technique for *regressus* means that these examples have to stand as such; study of them will show at most the possible causes of the degeneration of the society in question. Hobbes's theorized state of nature will offer a demonstration *propter quid* of the complete dissolution, giving the actual cause—both formal and efficient—of the natural condition. *Leviathan* thus not only offers images for those so inclined to the analytic art and to inferring the causes of the state of nature from historical examples but also, and more important, offers a synthetic account of what the state of nature would look like, from its causes. By clearly separating the historical examples, which are demoted to rhetorical effect, from the theoretically constructed natural state, Hobbes effects a solution to the Machiavellian problem of exemplarity: the examples do instruct, in the sense that they appeal to a reader's sentiments, but their failure to produce a principle for imitation is no longer relevant, since they have been barred from doing so from the start.[6]

Since the natural condition is, even when considered hypothetically, a historical condition, it is appropriate to begin with Hobbes's understanding of historical writing, as it is not unreasonable to suppose that what he praises in historical writing is something he then tries to implement in the historical sections of *Leviathan*. There, when he divides the subjects of knowledge, he relegates history to the knowledge of fact, as opposed to the "conditionall" knowledge (if x, then y) required of philosophy (L 9.1–2, 60). He offers several clues to understanding this sort of schema in his introduction to his Thucydides translation. He notes that two general aspects of Thucydides' writings should be considered: their truth and their elocution. Good history requires both, for "the latter without the former is but a picture of history; and the former without the latter, unapt to instruct" (EW VIII, xx). This clear separation resolves the ambiguity in a Machiavellian account: the eloquent adduction of examples is explicitly said to be unable to instruct. Hobbes takes the truth of Thucydides' narrative as established beyond doubt, and "no man hath ever yet called it into question" (*ibid.*). For the elocution there are also two aspects to consider: disposition or method, and style. The virtue of Thucydides' disposition is that he follows "distinctly and purely the order of time throughout," such that "the grounds and motives of every action he setteth down before the action itself" either narratively or through contrived deliberative orations.[7] Thucydides does judge actions, but he avoids digressions. In sum,

> As having so clearly set before men's eyes the ways and events of good and evil counsels, that the narration itself doth secretly instruct the reader, and more effectually than can possibly be done by precept. (EW VIII, xxii)

In other words, the concise use of images and a linear narrative structure are as effective in inducing the belief in the reader's mind as scientific or philosophic demonstration. I will return to this point; it is worth mentioning here, however, that in his later mathematics, Hobbes says exactly the same thing about analysis: it is able to induce a psychological condition comparable to demonstration while nonetheless remaining nonscientific (OL IV, 39). As Hobbes quips to Bellarmine, "examples *prove* nothing" (L 42.135, 402; my emphasis; this is repeated verbatim at OL III, 433). In referring to Thucydides' style, Hobbes underlines the point. He invokes Plutarch, who says that "these things . . . are so described and so evidently set before our eyes, that the mind of the reader is no less affected therewith than if he had been present in the actions" (EW VIII, xii). It is this standard to which Hobbes's examples of disorder in *Leviathan* strive. *Leviathan*'s remarks on the virtues of a history underscore the point: it requires both fancy (good imagination—*phantasia*—cf. OL III, 55) and judgment, with the latter dominant, "because the goodnesse consisteth, in the Method, in the Truth, and in the Choyse of the actions that are most profitable to be known." Fancy is demoted to adornment (L 8.5, 51).

In *Leviathan*, when Hobbes underscores that his state of nature is hypothetical, conceding the objection that "there was never such a time, nor condition of warre as this," he nonetheless offers three examples "where they live so now" (L 13.11, 89). One is of Native Americans, another is of life during wartime, and the final is of the relations between nations.[8] These examples are initially troubling and have been the source of much confusion, since they can imply that Hobbes does, in fact, think that the state of nature exists. However, none of them withstands scrutiny as actualizing the state of nature. I will treat them in reverse order. International relations, Hobbes says, is conducted under a "posture of War," as is evidenced by the arms buildups and posturing of sovereign authorities against one another. However, he explicitly disqualifies this international anarchy from being equivalent to the state of nature: "because they [sovereigns] uphold thereby, the Industry of their Subjects; there does not follow from it, that misery, which accompanies the Liberty of particular men" (L 13.12, 90).[9]

The case of civil war is more complicated, and the English Civil War would have been an event immediately on the minds of Hobbes's readers. He is quick to draw their attention to it:

> It may be perceived, what manner of life there would be, were there were no common Power to feare; by the manner of life, which men that have formerly lived under a peacefull government, use to degenerate into, in a civill Warre. (L 13.11, 89–90)

No English reader of 1651 could have missed the image, and *Leviathan* reinforces it with other, scattered references to the Civil War. The parallel passage in the Latin edition also refers to civil war, though more briefly, and adds a reference to Cain and Abel (OL III, 101).[10]

That said, the English Civil War could not have been the *bellum omnium*. Evidence for this is both textual and conceptual. I will return to the conceptual argument below, with reference to Hobbes's rejection of *regressus* technique. The textual evidence is not unequivocal, but there are several instances of it. First, Hobbes does not say that the Civil War *was* a state of nature. He says that it might help his readers to perceive what life in the state of nature might be like, not what the state of nature was. Second, the distinction is supported elsewhere in the text. For example, he claims that the disadvantages of any form of government are "scarce sensible, in respect to the miseries, and horrible calamities, that accompany a Civill Warre; or that dissolute that dissolute condition of masterlesse men, without subjection to Lawes, and a coërcive Power to tye their hands from rapine, and revenge" (L 18.20, 128). The "or" is indeed not entirely clear, but the Latin edition supports reading it conjunctively, rather than equivocally. There, Hobbes suggests that the inconveniences are scarcely sensible compared to "the calamities of war, and which are contained in the natural condition of man" (OL III, 140). Third, when Hobbes does speak more of the English Civil War, he presents it as a war of factions, not a complete dissolution. Thus, the division of the rights of the sovereignty was a necessary condition of the war:

> For unlesse this division precede, division into opposite Armies can never happen. If there had not first been an opinion received of the greatest

> part of *England*, that these powers were divided between the King, and the Lords, and the House of Commons, the people had never been divided, and fallen into this Civill Warre; first between those that disagreed in Politiques; and after between the Dissenters about the liberty of Religion; which have so instructed men in this point of Soveraign Right, that there be few now (in *England*,) that do not see, that these rights are inseparable, and will be so generally acknowledged, at the next return of Peace; and so continue, till their miseries are forgotten; and no longer, except the vulgar be better taught than they have hitherto been. (L 18.16, 127)

This analysis recurs in much greater detail in his considered treatment of the Civil War, *Behemoth*. Again, Hobbes repeatedly shies away from characterizing the Civil War as a *bellum omnium*: the Rump parliament was an "oligarchy" (1990, p. 156), by 1653 Cromwell had the "supreme strength" (1990, p. 180), and so forth. Even before Cromwell was ascendant, in 1646, Hobbes maintains that "the right [of political authority] was certainly in the King, yet the exercise was yet in nobody" (1990, p. 135). Hobbes presents the war as one of competing factions, rather than a dissolute mob of individuals. Not only that, but also members of these factions act according to what they have been normed to do: those taught seditious doctrines by the universities or Presbyterians behave seditiously—none of this in keeping with the state of nature.

Hobbes thinks that the Civil War rather happened because of the existence of a rival center of power in parliament, and his full consideration of it intends to illustrate the need for a unitary executive, not to illustrate the state of nature. The rival divisions of power have rival material bases, and the characters in the dialogue emphasize that this is to be a general theory of civil war:

> B: It seems, not only by this, but also by many examples in history, that there can hardly arise a long or dangerous rebellion, that has not some such overgrown city [London], with an army or two in its belly to foment it.
>
> A: Nay more; those great capital cities, when rebellion is upon pretence of grievances, must needs be of the rebel party: because the grievances are but taxes, to which citizens, that is, merchants, whose profession is their private gain, are naturally mortal enemies; their only glory being to grow excessively rich by the wisdom of buying and selling. (126)

Behemoth thus answers to the same sort of view of historical narrative that Hobbes ascribes to Thucydides.

Against those who study the Greeks to learn the benefits of democracy, Thucydides' *History* also provides further examples of the disintegration of societies into civil war, again along factional lines. During the Corcyraean sedition, when confined to the city by Athenian galleys:

> The Corcyraeans did nothing but kill such of their city as they took to be their enemies. . . . All forms of death were then seen; and . . . whatsoever had happened at any time, happened also then, and more. For the father slew his son; men were dragged out of the temples, and then slain hard by; and some immured in the temple of Bacchus, died within it. (III.81; EW VIII, 347)[11]

Thucydides adds a harrowing description of a time in which, as he summarizes, "all Greece . . . was in commotion." Specifically, "quarrels arose everywhere between the patrons of the commons, that sought to bring in the Athenians, and the few, that desired to bring in the Lacedaemonians" (III.82; EW VIII, 347). Thucydides describes a complete implosion of civil society, in which (for example), "to be revenged was in more request than never to have received injury" (III.83; EW VIII, 347). He is, however, very clear that this is the consequence of an organized factionalism, as "the cause of all this is *desire of rule*, out of *avarice and ambition*; and the zeal of contention from those two proceeding." The public was treated like a prize, its members disposable parts in the machinery of elite ambition:

> For such as were of authority in the cities, both of the one and the other faction, preferring under decent titles, on the political equality of the multitude, the other the moderate aristocracy; though in words they seemed to be servants of the public, they made it in effect but the prize of their contention: and striving by whatsoever means to overcome, both ventured on most horrible outrages, and prosecuted their revenges still farther, without any regard of justice or the public good, but limiting them, each faction, by their own appetite: and stood ready, whether by unjust sentence, or with their own hands, when they should get power, to satisfy their present spite. (III.82; EW VIII, 350)

Thucydides had noted that linguistic chaos accompanied this condition, an observation that had been applied by thinkers like Montaigne to the late Renaissance.[12] Undoubtedly, this sort of description was also on Hobbes's mind when he declared that a failure to recognize equality was a mark of pride, and when he carefully tried to insulate the monarch from any sort of private interest (indeed, this is part of what monarchy is superior to both aristocracy and democracy: the monarch has no possible factional interest). It presents a scenario very similar to the one Hobbes diagnoses in the English Civil War, and indeed provides evidence of what life would be like under the state of nature. Part of why the account is so harrowing is that this is a manipulated natural state, the result of fighting among only two factions. The imagination is led to contemplate: what would life be like if the war were truly of *omnium contra omnes*? Civil wars teach that more than one center of power is catastrophic; the catastrophe of their being literally no common power would be, this implies, complete. Civil wars end; however, the state of nature stipulates that such an end is impossible, as "a war which cannot be brought to an end by victory because of the equality of the contestants is by its nature perpetual" (DC I.13).

Hobbes's final illustration of what life would be like in the state of nature is of Native Americans. In *Leviathan*, he remarks that the "savage people in many places of America, except the government of small Families" would otherwise have "no government" (L 13.11, 89–90). Hobbes invokes Native Americans in other texts as well: in *De Cive*, "the present century presents an example" of perpetual war "in the Americas" (DC I.13); and in *Elements of Law* refers to "the experience of savage nations that live at this day" (EL 14.12). This view exploits a tension in prevailing seventeenth-century English views of the Americans, who were held to be both bestial and peaceful, "a loving, a very true, and just dealing people," as Sir Francis Drake suggested, despite their supposedly inferior culture.[13] The effect of Hobbes's reference was to equate anarchy with savagery; as Richard Ashcraft puts it, Hobbes "fused together his contemporaries' cultural prejudices and their fear of a return to civil war and anarchy into an effective ideological picture of anarchy as savagery" (1971, p. 1108).

This is indeed the rhetorical effect. But the Hobbesian Native Americans are not fully inhabitants of the state of nature. As the *Leviathan* passage suggests, they have the rule of small families. The Latin version explicitly suggests that this is a *de minimis* level of law: they "are subjected to paternal laws through small families" (OL III, 101).[14] In Hobbes's mature theory, families are political, not natural entities.[15] Hobbes even claims, improbably, that the dominion parents have over children is "from the Childs Consent, either expresse, or by other sufficient arguments declared" (L 20.4, 139). Then, invoking the principle that those in the state of nature should be understood as equal, he rejects patriarchal theories that assign natural dominion to the father: "whereas some have attributed the Dominion to the Man onely, as being of the more excellent Sex; they misreckon in it.

For there is not alwayes that difference of strength, or prudence between the man and the woman, as that the right can be determined without War" (*ibid.*). Reason thus dictates that they treat one another as equals. In the condition of "meer nature," without civil laws to govern the right, then, parents either "between themselves dispose of the dominion over the Child by Contract; or do not dispose thereof at all." He then notes a default: "if there be no Contract, the Dominion is in the Mother," since (a) paternity cannot be known without declaration by the mother as to who the father is, and (b) the infant owes its life to the mother. In developing the second of these points, Hobbes offers what I take to be an explanation of how infants can contract: "every man is supposed to promise obedience, to him, in whose power it is to save, or destroy him" (L 20.5, 140). Assigning the infant to the mother by contract is a way of invoking theoretical principle: not a principle of consent but a principle of status, as indicated by the supposition. A child may or may not like its parents—but insofar as it depends on them for its continued existence, we are to suppose that they have right over it. The rule of families, then, establishes that Native Americans have a political system of sorts.

Hobbes's description of the Native Americans is similar to his rendering of what Thucydides calls "the imbecility of ancient times" (I.3; EW VIII, 3). Thucydides describes a period of nomadism and forced population migrations. As Hobbes translates the relevant passage,

> At first there were often removals, every one easily leaving the place of his abode to the violence always of some greater number. For whilst traffic was not, nor mutual intercourse but with fear, neither by sea nor land; and every man so husbanded the ground as but barely to live upon it, without any stock of riches, and planted nothing; (because it was uncertain when another should invade them and carry them all away, especially not having the defence of walls); but made account to be masters, in any place, of such necessary sustenance as might serve them from day to day. (I.2; EW VIII, 2)[16]

Thucydides proceeds to show that differences in soil fertility contributed to these dislocations, with places having the most fertile soil being the least stable. Athens, then, "from great antiquity for the sterility of the soil free from seditions, hath been inhabited ever by the same people" (I.2; EW VIII, 3). Growing by an influx of war refugees, and unable to feed itself, Athens soon sent out colonies into Ionia. The two most striking parallels between the Hobbesian state of nature and the account of ancient Athens here are the lack of commerce and subsistence-level existence, on the one hand, and the radical uncertainty which generates this condition, on the other. As I will argue, one feature of the Hobbesian state of nature is precisely this uncertainty. Again, here, however, the ancient Greeks provide an example of what life might be like in the natural state, but it is not a generalized state of nature: there is geographic diversity which serves partly to explain the conflict, some people have evidently succeeded in forming alliances such that they are able to evict smaller groups of people (thus denying the war of all against all), and Athens remains as a stable place of refuge. At the very least—and this is the important point—such features make it possible to doubt whether this was in fact the state of nature and what might be its causes.

Study of the available historical and anthropological evidence, then, produces a number of clues as to what the state of nature would be like. Life would be "nasty, poore, solitary, brutish and short," something like it happens during a civil war or for Native Americans. One of the principal problems would be lack of commerce for commodious living; one of the reasons for this lack is insecurity. This insecurity is generated, *inter alia*, by the lack of commonly accepted meanings for moral terms. The lack of linguistic certainty both radicalizes and serves as a linchpin for all of the other uncertainties. As I have suggested, these images and resonances together (and there are others: commentators have pointed to sources ranging from Calvinistic descriptions of the Biblical fall to Lucretius) paint a compelling picture of why one would not want to live in the state of nature. Although the textual evidence is admittedly not decisive, I thus think there are very good reasons to doubt that any one of them was an exact archetype for the Hobbesian natural state.

To these textual reasons, I want to add a conceptual one, which I do take to be decisive.[17] As I indicated in chapter 3, Hobbes heavily prioritizes synthetic over analytic reasoning. The most direct evidence of this is that he specifically rejects any theoretical solution of the scholastic *regressus* problem, which retained the scientificity of demonstrations *quia* (from effects) by providing a technique for moving between them into demonstrations *propter quid* (from causes). In this context, it is easy to see why historical examples never suffice to generate the state of nature: the most such a demonstration could convey is that a given historical situation, with a manner of life similar to what the state of nature would be like, *might possibly* be caused by its being the state of nature. Hobbes says exactly this of the study of civil wars:

> For he that hath seen by what courses and degrees, a flourishing State hath first come into civil warre, and then to ruine; upon the sight of the ruines of any other State, will guesse, the like warre, and the like courses

> have been there also. But this conjecture, has the same incertainty almost with the conjecture of the Future; both being grounded onely upon experience. (L 3.10, 23)

To conclude that any given civil war or other comparable situation instantiated the state of nature would be to depart science and enter into the realm of speculation.

Hobbes specifically repositions moral philosophy such that it would conform with the principles outlined here. The laws of nature, as he indicates, are those which, if followed, would conduce to peace. "The true and onely Moral Philosophy" is then the science of these laws (L 15.40, 110). Thus, moral philosophy is concerned with the study of the means by which peace can be achieved, or, alternately, of the causes of its breakdown. There would be two techniques for the study of moral philosophy. The first is analytic and would proceed from actual examples of virtue or peace. Such is the procedure followed by Machiavelli, and (according to Hobbes) by the ancients more generally. This is also apparently the sense of moral philosophy indicated on Hobbes's division of the subjects of knowledge, where ethics is the study of the consequences of the passions of people (L 9 chart, 61), and ultimately therefore a part of natural philosophy. We know already that natural philosophy is not subject to demonstration: Hobbes's caution to this effect, as I have already indicated, is that "because of natural bodies we know not the construction, but seek it from the effects, there lies no demonstration of what the causes be we seek for, but only of what they may be."[18] Speculation about civil war and the natural state of man and whether a given historical situation is really such a "natural state" may be compelling, but "*scientia non sit*," as Hobbes says of demonstrations *quia* generally. Hobbes indicts the ancient moral philosophers, and distinguishes his own procedure, on precisely these grounds:

> The science of Vertue and Vice, is Morall Philosophie; and therefore the true Doctrine of the Lawes of Nature, is the true Morall Philosophie. But the Writers of Morall Philosophie, though they acknowledge the same Vertues and Vices; yet not seeing wherein consisted their Goodnesse; nor that they come to be praised, as the meanes of peaceable, sociable, and comfortable living; place them in a mediocrity of passions: as if not the Cause, but the Degree of daring, made Fortitude; or not the Cause, but the Quantity of a gift, made liberality. (L 15.40, 111)[19]

From this point of view, the moral lessons that one draws from historical examples are useful in the sense that they are good at getting people to do things, but they are not scientifically valid, and in that sense, not even properly dispositive for action. So too, the Machiavellian reversal of Aristotle, according to which one is to value extremes of action over the middle, presents the flip side of the same coin: from a Hobbesian vantage point, it is no surprise that Machiavelli's principles are underivable and unstable. On the contrary, the derivation of the state of nature in the *Leviathan* is intended to be synthetic; whatever rhetorical or textual slippages the argument might introduce, Hobbes's intent is clearly to provide a demonstration of the infelicity of the state of nature by its causes, and these causes are to be stipulated as part of the experiment. That any given state of affairs might or might not look like this is a contingent fact.

5.2 Contractual Failure

If there is no empirical model from which we can determine the contours of the natural state, what are they? In this section, I will look at a surface-level feature of Hobbes's natural state: his claim that those in it do not form contracts. In the following sections, I will attempt a theoretical reconstruction, the point of which will be to show that the disorder of the multitude is, in Hobbesian terms, demonstrable. Remarks by two of Hobbes's contemporaries can serve to frame the discussion. First, Eachard's "reconsideration" will try to show that Hobbes has attributed his own bad nature to the entire human condition and that if one looks in his "Logik," one will "find a whole *Book* full of nothing but *new words*" (1958, p. 18), and that to "have *Books* tailed together by far fetched contrivances; and to swagger them off for *demonstrations* . . . is so very idle" (1958, p. 34).[20] For his part, Clarendon complains of "Mr. *Hobbes*'s Logic, which is a great presumption, that from very true Propositions he deduces very erroneous and absurd Conclusions" (1676, p. 41). Specifically, Hobbes "takes many things for granted which are not true; as that *Nature hath made all men equal in the faculties of body and mind*" (1676, p. 26). This misreading underscores the extent that one should focus on Hobbes's "logick." To recall, the relevant passages in *Elements of Law*, in which we are to contemplate a sovereignty which is "like a creation out of nothing by human wit" (EL 20.1), is that "men considered in mere nature, ought to admit amongst themselves equality" (EL 14.2).[21]

As Hobbes famously says, those in the state of nature do not form covenants. Since the making of covenants is both the originary and paradigmatic case of a political act, and since—as will be evident—that act is irrational in the state of nature, the impossibility of covenants in the state of nature is the impossibility of political rationality in it. He suggests but does not develop the point in *Elements of Law*, where he notes that covenants are impossible in the state of nature but that "after the introduction of policy and laws, the case may alter" (EL 16.13). His fuller discussion of the issue is in *Leviathan*, where he challenges the Biblical fool who "says in his heart that there is no God." The same fool "hath sayd in his heart, there is no such thing as Justice," that one may "make, or not make; keep or not keep Covenants" and that such practice "was not against Reason, when it conduced to ones benefit" (L 15.4, 101).[22] Initial guidance on how to read this passage can be found by noting that the argument is a seventeenth-century commonplace, with the position of the fool usually taken by Carneades. Carneades was a "symbolic hero" to classical rhetors, famous both for skepticism and for his having successfully argued for justice on one day and against it on the next.[23] Both Grotius, before Hobbes, and Locke after him, let Carneades espouse the reduction of justice to expediency. As Grotius cites the view, "all creatures, men as well as animals, are impelled by nature toward ends advantageous to themselves; that, consequently, there is no justice" (1962, *Prol.*V.2).[24]

Carneades' specific discussion of justice is reported in Lactantius' *Divine Institutes* V.14.[25] There, the claim is not simply that there is no justice but that the just person is a fool. Carneades cites three kinds of examples. First, great conquering nations would have to give back everything they took and reduce themselves to poverty in order to achieve justice. Second, the just person acts against his commercial interest: someone who discloses a defect in a house before selling it is just but unwise, since he will sell it either at a bad price or not at all. Third, the just person puts himself into mortal danger: a just person would not seize the only available flotsam after a shipwreck, but a wise one would. Carneades thus poses two questions. First, whether a just person is foolish, and second, whether the natural drive to expediency *eo ipso* entails that justice does not exist.

Hobbes's answer to the first question is to split it into questions of the law of nature *in foro interno* and *in foro externo*. In the state of nature, the laws of nature bind to a desire that they take place but no more; whereas, in civil society, they are actually binding on performance (L 15.36, 110). This split allows Hobbes to realign justice with expediency. In the state of nature, it is instrumentally rational to desire peace, because it is a necessary condition for civil society that its members desire peace. Since civil society is rationally desirable as an end, the necessary conditions for its existence are also rationally desirable. The sufficient condition—that all the members of the state of nature perceive each other as genuinely desiring peace—would of course probably never occur. But this contingent fact does not disturb the original argument. On the other hand, it would be irrational to perform the laws of nature in the state of nature, as Hobbes makes immediately clear. Anyone who was the only person to perform the laws of nature would bring immediate ruin upon himself, which violates the law of self-preservation (L 15.36, 110). As he explains slightly earlier:

> For as long as every man holdeth this Right, of doing any thing he liketh; so long are all men in the condition of Warre. But if other men will not lay down their Right, as well as he; then there is no Reason for any one, to divest himself of his: For that were to expose himself to Prey, (which no man is bound to) rather than to dispose himselfe to Peace. (L 14.5, 92)

For this reason, Hobbes concludes, there can be no justice or injustice in the state of nature. Specifically, justice can occur only in the performance of contracts and covenants. In the state of nature such covenants are irrational, and so there is no justice in the state of nature. Indeed, the "Cardinall vertues" in the state of nature are force and fraud (L 13.13, 90).

The introduction of a "common power" changes the calculation by imposing sufficient penalties for nonperformance that a rational person will not just desire that the laws of nature bind but actually be willing to submit to them. In the paragraph before the fool makes his appearance, Hobbes argues that there needs to be "some coërcive Power, to compell men equally to the performance of their Covenants, by the terrour of some punishment, greater than the benefit they expect by the breach of their Covenant" (L 15.3, 100–1). Instrumental rationality thus dictates that one follow one's covenants: by imposing a penalty for nonperformance greater than any possible short-term gains from it, the governmental power suddenly makes it rational to perform covenants. At the point it is transparently rational for actors to perform covenants, it becomes rational to enter them; at the point it is rational to enter covenants, it is rational to follow the laws of nature *in foro externo.*[26] As Hobbes outlines the various remaining laws of nature, most of them have to do with ensuring precisely this transparent contract performance, and the breach of most of them he puts in terms of a desire for war. The commonwealth thus prescribes two penalties for nonperformance. The first is whatever is allowed by its laws, and the second is that nonperformance increases the likelihood of the commonwealth ending and a relapse into the state of nature. Someone who "having sufficient Security, that others shall observe the same Lawes towards him, observes them not himselfe, seeketh not Peace, but War; & consequently the destruction of his Nature by Violence" (L 15.36, 110). This nuclear option overwhelms all other calculation: by stipulation, the fool (and this, presumably, is why Hobbes cites the fool and not Carneades) who denies the existence of justice is an atheist and will be unmoved by threats of eternal damnation.[27] Carneades' second argument is thus met: in any situation where being "just" would be being foolish, all that being just requires is the desire that justice happen. Any further extension of the term is meaningless.

The first argument, that self-preservation *eo ipso* undermines justice, is addressed by the state of nature/commonwealth distinction in combination with Hobbes's nominalist commitments. In his answer to Carneades, Grotius cites animals which defer their own good for their offspring and suggests that people possess "an impelling desire for [peaceful] society" (1962, *Prol.* VI.2). Hobbes can already avail himself of the same argument: all rational people *desire* society; it just happens to be irrational to do anything about that desire unless the performance of others can be guaranteed. This response of course answers the letter but not the spirit of the objection. Natural law, in the sense that Carneades thinks he is refuting, would be some sort of eternal dictate mandating certain conduct. Hobbes straightforwardly announces that his laws of nature fit the description. Because they are derived from reason, the laws of nature "are Immutable and Eternall; for Injustice . . . and the rest, can never be made lawfull." This is because "it can never be that Warre shall preserve life, and Peace destroy it" (L 15.38, 110). Hobbes even goes as far as to claim that pursuit of narrow self-interest is not a sufficient condition for describing a person as just: in the case of an unjust person, his "Will is not framed by the Justice, but by the apparent benefit of what he is to do" (L 15.10, 104).

That this response seems intuitively unsatisfactory—"justice" seems here to mean "benefit" in the larger sense and "benefit" to mean "benefit in the short-term"—suggests that the real complaint is that basing everything on self-interest takes the lawgiving authority away from a divine power.[28] Hobbes deploys two definitions to address this problem. First, he says, the "laws of nature" are not strictly speaking laws at all, since law requires "the word of him, that by right hath command over others," whereas the laws of nature "are but Conclusions, or Theorems concerning what conduceth to the conservation and defense of themselves." Thus, the command from reason to self-preservation does not affect the existence of law *per se* at all. The only meaningful sense of the law of nature as law is as the command of God, as Hobbes explains in the next sentence: "if we consider the same Theormes, as delivered in the word of God, that by right commandeth all things; then are they properly called Lawes." Hobbes's nominalism establishes how this is to be understood. Since the theorems as delivered "in the word of God" require interpretation, and since interpretation requires a sovereign in order not to devolve into war, the content of these laws is determinable only in a commonwealth and by the sovereign. Hence, it is meaningless to speak of the word of God in a natural state. Hobbes's other definitional strategy achieves the same result, but directly: by defining justice in terms of covenants, and arguing that covenants will not happen in the state of nature, he again makes discussions of justice in the state of nature meaningless; in the war of all against all, "the notions of Right and Wrong, Justice and Injustice have there no place" (L 13.13, 90).[29] Hence, "nothing can be Unjust" (*ibid.*)—not because justice does not exist but because it is a category error to apply the term to the natural condition.

Hobbes's specific answer to the fool underscores all of these points and further emphasizes the impossibility of covenants (promises of mutual future performance) in the state of nature. How to read this answer has been the source of considerable confusion, a confusion abetted by his imprecise formulation:

> Not of promises mutually, where there is no security of performance on either side; as when there is no Civill Power erected over the parties promising; for such promises are no Covenants: But either where one of the parties has performed already; or where there is a Power to make him performe; there is the question whether it be against reason, that is, against the benefit of the other to performe. (L 15.5, 102; emphasis added)

As Pasquale Pasquino (2001) has observed, the "or" renders the passage ambiguous, and suggests the possibility of people in the state of nature forming a contract with one another. Despite this grounding, the reading seems not to be generally supportable. On Hobbes's account, practical rationality in the state of nature is impossible, which means that for him, political subjectivity is an effect of society, not its cause. First, as Pasquino points out, the parallel passage in the Latin revision of *Leviathan* replaces the "or" with an unambiguous "and," clearly indicating that the discussion is about contracts in civil society.[30] More important, the civil-society reading coheres much better with Hobbes's discussion of covenants more generally.

In chapter 14, Hobbes writes: "if a Covenant be made, wherein neither of the parties performe presently, but trust one another; in the condition of meer Nature, (which is a condition of Warre of every man against every man,) upon any reasonable suspition, it is Voyd" (L 14.18, 96). As will become evident, this is tantamount to saying that no covenant is possible in the state of nature because the required fulfillment conditions cannot obtain. On the one hand, since Hobbes stipulates that trustworthy people are a minority in the state of nature, a prudent person would wait to trust someone with a good reputation. However, as I will argue in detail below, the conditions for developing a stable reputation do not exist, so a prudent person would trust no one. On the other hand, and correlatively, *all* suspicion is "reasonable" in such a condition. Indeed, reason itself will turn out to be indistinguishable from such suspicion, a reading that is supported with reference to the Latin text, which drops the reasonability requirement: "this pact in the condition of mere nature, that is, in war, if any sort of suspicion about performance intervenes [*si quaecunque interveniat suspicio de praestando*], is invalid" (OL III, 107). It should be noted that the underlying problem is semiotic, as evidenced by the specification of "contract." A contract is a "mutuall transferring of Right" and can be either explicitly or implicitly declared. Express "signes of Contract . . . are words spoken with understanding of what they signifie." Implicit signs of contract require something which "sufficiently argues the will of the Contractor" (L 14.9, 14.14, 94).[31] A contract, then, requires a sign of some sort. As I will indicate, neither form of sign will be available in the state of nature because the state of nature stands for the impossibility of semiotics.

5.3 Semiotic Meltdown

Hobbes isolates three primary causes of war in the natural state: competition for natural resources and other rivalrous goods, diffidence ("constant despayre," L 6.20, 41; despair is appetite without hope of attaining its object, L 6.15, 41) and the desire for glory (L 13.6, 88). The interaction of these factors, on Hobbesian grounds, is sufficient to produce, if not constant active fighting, at least "a tract of time, wherein the Will to contend by Battell is sufficiently known." The condition is analogous to bad weather, the nature of which "lyeth not in a shore or two of rain; but in an inclination thereto of many dayes together" (L 13.8, 88).[32] It is a period of ominous foreboding, of individuals and small groups in distrustful and ephemeral interactions, all of which are overdetermined by the very real possibility of devolution into deadly, anarchic violence from which no one can reasonably feel secure. There has been much discussion of whether or not individuals could form voluntary associations in the natural condition and thereby bootstrap themselves out of it.[33] One element which has been consistently underemphasized in this context is what one might call a semiotic meltdown. That is, a central difficulty in the state of nature is the lack of a stable system of signification.[34] This absence pervades all aspects of a description of the natural state and needs to be underscored in any complete analysis. I will discuss this semiotic condition in some detail and then indicate how it exacerbates all three of the causes of war that Hobbes analyzes. The sovereign, as I will indicate, is charged with remedying these factors. That successful use of language is the *sine qua non* for society is indicated in the beginning of the chapter on speech, where Hobbes says that without the invention of speech, "there had been amongst men, neither Commonwealth, nor Society, nor Contract, nor Peace, no more than amongst Lyons, Bears, and Wolves" (L 4.1, 24). That there is a constant risk of semiotic failure is emphasized a few paragraphs later, when the four abuses of speech all derive from imprecise and unstable usage (L 4.4, 25–6).

Hobbes sets the stage for his discussion of the natural state by specifying several antecedent conditions. First, there is no survivability. This is a defining condition, and Hobbes remarks at the outset of the chapter that "as to the strength of body, the weakest has strength enough to kill the strongest" (L 13.1, 87). People will either realize this fact about themselves or not. Those who do realize it will be aware that they have to act without the presence of a reliable deterrent against the actions of others; this means that they do not have the luxury of waiting to see what another's actual intentions are. Those who do not realize it will tend to behave recklessly as they overestimate their own strength. Whether or not people realize this about themselves, they will certainly realize it about others and enter conflicts assuming that others have no reliable ability to deter them.

Second, people are heavily driven by their passions. These passions may or may not be sufficient by themselves to generate the *bellum omnium*, but they are certainly sufficient to be a significant factor in its emergence.[35] As I indicated in the previous chapter, desire is constitutive of human life for Hobbes, and this is the reason why one can posit a "general inclination of all mankind, a perpetuall and restlesse desire of Power after power, that ceaseth onely in Death." It is not just that people might or might not be greedy; even a modest person "cannot assure the power and means to live well, which he hath present, without the acquisition of more" (L 11.2, 70). Not only that, but also humans are prone to vain glory, the overestimation of their own abilities and the underestimation of others, a condition that promotes reckless behavior—"ostentation" or "rash engaging" (L 11.10–11, 72). As Hobbes puts it elsewhere, "the Passions that most of all cause the differences of Wit" may all be reduced to "Desire of Power" (L 8.15, 53); and, "the Passion, whose violence, or continuance maketh Madnesse, is either great *vaine-Glory*; which is commonly called *Pride*, and *self-conceipt*; or great *Dejection* of mind" (L 8.18, 54).[36] Those who have vain glory become incapable of rational behavior, as do those who despair of their lives in the natural condition. More generally, strong passions blur the line between rational and irrational behavior, and "if the Excesses [of the passions] be madnesse, there is no doubt but the Passions themselves, when they tend to Evill, are degrees of the same" (L 8.20, 54). The somewhat later verdict of *De Homine* is bluntly that the passions "frequently obstruct right reasoning" because they induce preferences for immediate and only apparent goods, at the expense of pursuit of actual, long-term goods (DH 12.1).

Third, conflicts over resources are inevitable, and it is to be assumed that resource scarcity is a fundamental feature of the natural condition, again by stipulation. Resource scarcity leads to conflict; since people are driven by their passions, they will contend for power. Hobbes suggests that competition over "Riches, Honour, Command, or other power enclineth to Contention, Enmity, and War: Because the way of one Competitor, to the attaining of his desire, is to kill, subdue, supplant, or repell the other" (L 11.2, 70).

Into this already unstable environment, Hobbes injects all the complications of an unreliable system of signification. That signification will be important is emphasized in Hobbes's discussion of power, where he remarks simply that, since the uniting of strengths is power, it follows that "Reputation of Power, is Power; because it draweth with it the adhaerence of those that need protection" (L 10.3, 62). He follows with several different kinds of reputation which augment power. Reputation, of course, depends on signification, and so Hobbes's understanding of the natural condition also depends on how signification in it would work or not work. For Hobbes, to be in the human condition is to be continually bombarded with sense data, linguistic signs, expressions which might or might not be intended to be meaningful, and so forth. This is one result of the *annihilatio mundi* experiment of *Elements of Law* and *De Corpore*, which is designed to show that phantasmata are ineliminable: take away the whole world, and one would still have a head full of such products of the imagination. Hobbes writes:

> We must remember and acknowledge that there be in our minds continually certain images or conceptions of the things without us, insomuch that if a man could be alive, and all the rest of the world annihilated, he should nevertheless retain the image thereof, and of all those things which he had seen and perceived in it. (EL 1.8; cf. DCo VII.1)[37]

This line of thought is concisely echoed in *Leviathan*, where Hobbes remarks that "after the object is removed, or the eye shut, wee still retain an image of the thing seen, though more obscure than when we see it." Imagination—to which the understanding reduces (see Chapter 4)—is therefore properly understood as "decaying sense" (L 2.2, 15), or, as the Latin further specifies, "failing sense, that is, diluted and vanished phantasms" (OL III, 8)[38]

Since all of us are embodied differently, and since all of us bring a unique history of bodily interactions to any given situation, all of us perceive the world somewhat differently: "for though the nature of what we conceive, be the same; yet the diversity of our reception of it, in respect of different constitutions of body, and prejudices of opinion, gives everything a tincture of our different passions" (L 4.24, 31).[39] Not only do we perceive differently and therefore experience different passions, but also the forms of speech by which we express those passions are unreliable; they are indeed expressions, "but certain signes they be not; because they may be used arbitrarily, whether they that use them, have such Passions or not" (L 6.56, 45–6; cf. L 11.19, 73). The difficulties of interpreting others' affects in the state of nature are thus almost insuperable, at least for topics that matter. Not only does Hobbes stipulate that moral words have no place in the natural state, but the difficulties in interpreting the passions of others makes meaningful assumptions about other persons impossible.[40] Hence, Hobbes follows the description of inconstancy in the passions with an example involving moral words:

> Therefore in reasoning, a man must take heed of words; which besides the signification what we imagine of their nature, have a signification also of the nature, disposition, and interest of the speaker; such as are the names of Vertues, and Vices; For one man calleth *Wisdome*, what another calleth *feare*; and one *cruelty*, what another *justice*; one *prodigality*, what another *magnanimity*; and one *gravity*, what another *stupidity*, &c. (L 4.24, 31)

In the state of nature, such caution becomes almost impossible. Indeed, in his discussion of prudence, Hobbes stipulates that the reason people have differing degrees of prudence is the "difference of Passions," which in turn derives "partly from the different Constitution of the body, and partly from different Education" (L 8.14, 53). This traces another line of breakdown: a radically unstable field of experience makes prudence difficult, which makes judgments about the intentions of others more difficult. Several specific points follow.

(a) Deception will be rampant in the state of nature. There are at least two independent reasons for this. On the one hand, ignorance of causes—the lack of science—makes people credulous, and such credulity, "because men love to be hearkened unto in company, disposeth them to lying." Thus, "ignorance it selfe without Malice, is able to make a man both to believe lyes, and tell them; and sometimes also to invent them" (L 11.23, 74). On the other hand, rational pursuit of interest also provides incentives to deceive. People will either realize their natural weakness or not. If they do not, they will be vainglorious and misrepresent their own power in order to deceive others into respecting them more. If not, they will perceive their weakness and attempt to remedy it by sending signals of strength. That weakness was a cause of deceptive behavior was an early modern commonplace. To the example of Machiavelli should be added Descartes' proof for why the *genie malin* could not be God: precisely, that God is omnipotent, but deception is a sign of weakness. Those who could simply get what they wanted would have no need to deceive. Since no one in the state of nature could realistically assume that she could simply get what she wanted, she will be forced to turn to "secret machination" (L 13.1, 87) to achieve her objectives by other means. Even in the context of oaths, Hobbes speaks of the "Glory, or pride, of appearing not to need to breake it" (L 14.31, 99). Rational actors in this state would therefore need some way to verify the signals that they received from others before they could act on them, especially those signals that invited cooperative action and the lowering of one's defenses. But such confidence building measures and verification regimes are extremely difficult to establish in the natural state.

(b) As I have already indicated, Hobbes stipulates that moral and political words like "right" and "wrong" and "good" and "evil" are meaningless outside the civil state. This makes oaths meaningless. As I discussed in the previous section, Hobbes specifically says that those in the state of nature cannot contract, absent the state to ensure second performance. The point I wish to make here is that the absence of meaning for moral words makes it impossible to trust someone's utterances, even assuming *arguendo* that this person was not being deceptive and that the utterance in question was designed to authenticate the person's sincerity. Hobbes suggests that there are only two powers that someone might fear enough to keep an oath: God and other people. The fear of other people is insufficient in the state of nature.[41] Hence, "all therefore that can be done between two men not subject to Civill Power, is to put one another to swear by the God he feareth" (L 14.31, 99). Hobbes promptly adds that an "Oath taken according to any other Form or Rite" than the swearer's own religion is "in vain; and no Oath" (L 14.32, 99–100). In the state of nature, how would one know that someone had undertaken a valid oath, the breaking of which he actually feared? On the one hand, there is no common system of moral and religious signification. On the other hand, there is the constant problem of an actor emulating the Machiavellian prince, who need only appear, but not be, devout. The only reason one might trust somebody, then, becomes untrustworthy. As Hobbes puts it in *De Cive*, "we cannot tell the good and the bad apart, hence even if there were fewer evil men than good men, good, decent people would still be saddled with the constant need to watch, distrust, anticipate and get the better of others, and to protect themselves by all possible means" (DC Pfc. 12).

(c) Not surprisingly, those in the state of nature would be heavily inclined to superstition and other irrational fears. As Hobbes says very early in the text, the fearful are inclined to apparitions and confusion about whether they are asleep or awake (L 2.7, 18).[42] Such fearfulness accounts for ancient religions. This fearfulness also makes the fearful easily manipulated, as he immediately notes (L 2.8, 18–19). Superstition thus afflicts both the weak and the strong. The weak will be prone to dejection, which "subjects a man to causelesse fears; which is a Madnesse commonly called MELANCHOLY" (L 8.20, 54). Thus increasingly and irrationally afraid of things around them, their analytic abilities are slowly eroded by "Prognostiques from Dreams, false Prophecies, and many other things depending thereon" (L 2.8, 19). This "*feare* of power invisible, feigned by the mind, or imagined from tales" is called religion when authorized by the sovereign; since there is no sovereign in the state of nature, religion is indistinguishable from superstition (L 6.36, 42). A general lack of knowledge of causes, combined with such fears, make the multitude uniquely susceptible to general panic:

> *Feare*, without the apprehension of why, or what, PANIQUE TERROR . . . Whereas in truth, there is alwayes in him that so feareth, first, some apprehension of the cause, though the rest run away by Example; every one supposing his fellow to know why. And therefore this Passion happens to none but in a throng, or multitude of people. (L 6.37, 42)

In other words, a given person fears something and imagines some cause.[43] The remainder, fearing the same thing but without analytic skills—they reason by example—simply assume that the first apprehension of the cause is correct. The multitude is thus particularly susceptible to the violent rage and madness induced by those who claim supernatural inspiration (cf. L 8.21, 54–5).

The strong will view the mob's susceptibility to panic as an opportunity, and in the state of nature, everyone would have an incentive to manipulate this fear and those passions to get what they want—this is one variant of the "secret machination" to which Hobbes alludes at the start of that chapter (L 13.1, 87).[44] Of course, manipulating people's irrational fears increases the overall amount of irrational behavior in the natural state and therefore increases the need for defensive posturing. It also makes the emergence of rogue actors—those whose behavior is fundamentally irrational—inevitable. The problem with rogue actors is that they cannot be rationally deterred, which means that preemptive strategies are increasingly necessary against them. Since there is no obvious way to tell the ordinarily superstitious from the manipulative from the rogues, such preemptive strategies will become increasingly pervasive. The sovereign is to put a stop to all such nonsense; "if he give away the government of Doctrines, men will be frighted into rebellion with the feare of Spirits" (L 18.16, 127).

(d) Finally, problems of authentication run deep enough that there is reason to believe that, in the state of nature, it would be difficult to speak meaningfully of "persons." The reasoning behind this claim is somewhat circuitous, and perhaps not decisive, but it is significant enough, I think, to count as a factor in the difficulties in signification in the natural state. A person, according to Hobbes, is

> He, whose words or actions are considered, either as his own, or as representing the words or actions of an other man, or of any other thing to whom they are attributed, whether Truly or by Fiction. (L 16.1, 111)

He then describes three types of persons: natural, when the words or actions are the person's own; artificial when they "are considered as representing the words an actions of another" (L 16.2, 111). A "fictitious" person is an entity that cannot author actions but which can have a representative that speaks for its interests. This category in Hobbes is very broad, and can include children, the insane, and various kinds of inanimate objects, such as bridges. Since the representatives of inanimate things or the irrational have to be authorized by some entity other than those represented, fictitious persons can only exist in civil society, against the background of a stable system of representation (L 16.9, 113).

In analyzing how interstate relations might be understood on Hobbes's model, according to which the sovereignty is a person, David Runciman suggests that sovereign which tries to find the agent behind another state's threats will be disappointed: "he will never discern there the person of the state. All he will find is a group of natural persons, who cannot as a group authorise the words he hears" (1997, p. 17). Here, I want to suggest that an analogous situation occurs for natural persons in the state of nature. This demolition of the presumptive unity of the natural person is the radical conclusion of Hobbes's initial assault on the Aristotelian assumption of meaningful natural differentiation and of the political animal. It also drives a final nail into the coffin of any sort of semiotic transparency in the state of nature.

Recall from Chapter 4 that the Hobbesian psychology systematically refuses to countenance any sort of metaphysical unity in human subjects. Sense data enter into the organism of a person, and are eventually translated into imaginations and those in turn are represented as thoughts. In a central passage in *De Homine* (cited in the previous chapter), Hobbes underlines that intellect reduces to imagination, and he repeatedly rejects unifying notions of intellectual faculty or free will. What, then, is a natural person but a collection of competing passions, located in the same biological organism, some of which will be stronger at any one time than another? Whichever passion is strongest is the one that will determine what the organism does next; as he puts it, "that which is really in us, is . . . onely Motion, caused by the action of externall objects, but in appearence [as they appear to us, rather than as they necessarily are]" (L 6.9, 40). Hobbes is consistent enough on this point to attempt to turn it into a virtue for ethics: when Bramhall complains that punishment will be useless in a universe without free will, Hobbes replies that the opposite is true: only in a universe where personhood is reduced to the passions is there any chance of punishment actually working, that is, of it actually affecting the decisions of those punished or those who see others punished.

This condition creates an epistemological problem. A few paragraphs earlier in *Leviathan*, Hobbes says that "because the constitution of a mans Body, is in continuall mutation; it is impossible that all the same things should alwayes cause in him the same Appetites, and Aversions" (L 6.6, 39). The attempt to know a person's appetites from the outside is thus somewhat like the situation in understanding nature but worse. In physics, we are incapable of demonstration because we do not know the causes of the motions we observe: this is the theoretical consequence of his rejection of analysis and of a technique for *regressus*. The situation in attempting to know the appetites of another is much worse: each person exists at an equilibrium such that certain motions will affect her in a certain way, but this equilibrium constantly changes from moment to moment; the external motions which affect a person only do so through their representation, and this internal translation of the motion into some sort of signification is opaque from the outside; and in the case of whatever the person finally says, it will be impossible to know if she represents or misrepresents her unknowable internal state. On the stabilizing assumption that people generally are alike, introspection suggests that they will attempt to preserve themselves. But even this only helps a bit, since in the state of nature we do not know what they take to be preservative.

Given all of this, when I confront someone in the Hobbesian natural state, I have no way of knowing what is behind that person's utterances. If I try to investigate, her utterances dissolve into a group of natural passions, which cannot as a group authorize the utterances I hear, in the sense that it is the equilibrium among them that is responsible, not any one of them. And although in those who are not completely insane, there is some sort of equilibrium, it is unknown to an interlocutor. The state serves the same function here as it does with the representation of inanimate objects: it authorizes something to speak on its behalf. In the case of natural persons, the state allows me to operate on the assumption that my interlocutor is motivated by her "awe" of the sovereign power and that her appetites are regulated accordingly. But in the state of nature—defined as the absence of this dominant power—I can no longer rationally make that assumption. In this precise sense, it is possible to suggest that there are no "people" in the state of nature, or, rather, there are people, but it is impossible to know anything relevant about them.

The sum of all of those confusions allows Hobbes to attribute the confusion of the multitude to the multiplication of chaotic centers of signification. He argues:

> And be there never so great a Multitude; yet if their actions be directed according to their particular judgments, and particular appetites, they can expect thereby no defence, nor protection, neither against a Common enemy, nor against the injuries of one another. For being distracted in opinions concerning the best use and application of their strength, they do not help, but hinder one another; and reduce their strength by mutuall opposition to nothing: whereby they are easily, not onely subdued by a very few that agree together; but also when there is no common enemy, they make warre upon each other, for their particular interests. (L 17.4, 118)

This scenario's credibility is both augmented and defined by the inability of signification in the state of nature. Hobbes's subsequent differentiation between a human and an animal natural state makes this clear (L 17.6–12). Almost all of the uniquely human characteristics—competition for honor, the human desire for eminence, the human urge to innovate in public affairs, the "art of words, by which some men can represent to others, that which is Good, in the likenesse of Evill," the human ability to distinguish injury from physical damage, and the artificiality of human agreement—can only be addressed by stabilizing the semiotic field in which politics occurs.

The semiotic meltdown makes all three of the *casus belli* in the state of nature worse. Competition is worse because the lack of stable society means that "there is no place for Industry; because the fruit thereof is uncertain." It also becomes impossible to know about individual issues of resource allocation, which means that the rational strategy for survival is to assume that all resources are rivalrous, since the failure to procure such resources as do exist can result in one's demise. One central function of the Leviathan is precisely to enable commerce and industry, thereby enabling people to live above the subsistence level. Hence the seventh right of the sovereign is in allocating and establishing rules for property (L 18.10, 125). An immediate effect is the removal of the insecurity induced by everyone having a right to everything; a secondary effect in the consequent elevation of living standards is a reduction in the need for conflicts over those resources.[45]

Diffidence is also a systematically more difficult problem: one does not know what others' intentions and abilities are, except for the general operating assumption that one is vulnerable to them. Their signals are ambiguous, significations for the relevant moral and political terms are unsettled, and so forth. The rational actor thus assumes an aggressive defensive posture in order to avoid being subject to the "secret machination[s]" that Hobbes assumes are sufficient to kill even the physically strong (L 13.1). In addition, one's underconfidence, and the inability to base that underconfidence on evidence, makes it more difficult for experience to yield prudence (L 13.2), which in turn makes it that much more difficult to enter into situations in any other than a defensive posture. The Leviathan's intervention here, then, is partly to secure the transparency of communications.

Finally, the desire for glory, which is from the outset a problem of signification, is all the more intractable in the absence of communicative stability. The specific problem is the presence of multiple, competing centers of signification, such that "every man looketh that his companion should value him, at the same rate he sets upon himselfe: And upon all signs of contempt" tries to extract "a greater value from his contemners" (L 13.5, 88).[46] Since glory is a function of moral and political words and of societally mediated values like honor, the lack of those functions makes it difficult to express honor and difficult to understand when someone is or is not being sufficiently respectful. Furthermore, honor is a function of injury—not physical damage, but perceived insult. When determinations of the grounds for injury are privatized, the possible number of injuries increases exponentially, and the ability to guard against injuring others declines accordingly. Two factors further exacerbate the problem. First, everyone thinks himself above average in wisdom (L 13.2, 87), which means that he will demand more signs of respect than his actual abilities would merit. Second, the remedy for problems of glory is to set an example for others to follow—but for this example to work would require precisely the missing framework of stable communication that is missing in the natural state.[47] The sovereign, then, is specifically charged with remedying this situation by overdetermining its semiotic conditions:

> Considering what values men are naturally apt to set upon themselves; what respect they look for from others; and how little they value other men; from whence continually arise amongst them, Emulation, Quarrells, Factions, and at last Warre, to the destroying of one another, and diminution of their strength against a Common Enemy; It is necessary that there be Lawes of Honour, and a publique rate of the worth of such men as have deserved, or are able to deserve well of the Commonwealth; and that there be force in the hands of some or other, to put those Lawes in execution. . . . To the Soveraign therefore it belongeth also to give titles of Honour; and to appoint what Order of place, and dignity, each man shall hold; and what signes of respect, in publique or private meetings, they shall give to one another. (L 18.15, 126)

The discussion of honor thus confirms what Hobbes has already made abundantly clear: absent a sovereignty to order the semiotic field, natural human conflicts become a generalized state of war. This is not a result derived from historical study but from synthetic reconstruction.

5.4 Deter Thy Neighbor

What, then, of the logic of life in the natural state? Could people get out of it? Because a good deal of attention has been given to this question, I think it is worth reviewing in light of the failure of communicative systems I have outlined above. As I have already indicated, I think the tendency to frame this question as a problem (how to get out of the state of nature) for which Hobbes must provide a solution is misplaced. In what follows, I want to suggest that the question is aporetic in the sense that the logic of Hobbes's description, and its overdetermination by communicative failures, produce a series of interpretive dilemmas which make it impossible for hypothetical actors in the natural condition to rationally make the move to collective rationality presupposed by any bootstrapping process. In other words, that actors in the state of nature cannot form contracts is symptomatic of larger problems in the natural condition. I thus offer the following analytic comments to carry to its completion the implications of the Hobbesian state of nature. In games-theoretic terms, the general situation has to be that of a single-play prisoner's dilemma. To put it differently, "non-cooperation is the equilibrium state."[48] Because no one can ensure his or her own survivability, and because there is no credible way of sorting out the signals of another, all such situations are high risk.[49] The magnification of uncertainty makes it very difficult to rationally make the leap of faith required to assume that the game is iterated, and, consequently, makes estimation of how many iterations the game might continue even harder. The situation also makes it difficult to assume that one could live another day to educate one's interlocutor about the benefits of cooperation. Even the most rational of actors has to assume that her interlocutor either correctly understands his own weakness or not. If he does, he will try to deceive her into thinking he is stronger. If he does not understand his weakness, he will also posture aggressively to compensate. Thus, one is likely to encounter almost nothing but aggressive posturing, which further makes confidence-building measures difficult. Hence, Hobbes suggests that, in a condition of "tumult," those who distrust their own wit enough that they do not "consult" but instead "strike first" are "better disposed to victory" (L 11.10, 72).

Another risk in the state of nature is accidental conflict.[50] There are at least three reasons this risk is particularly high. First, the lack of survivability forces one's second-order decision strategy—the sorts of decisions one makes about how to behave in a crisis—to a much less stable posture. Is the gentleman approaching with a large axe hostile or cooperative? If I do not think that I can survive a hostile encounter, I will have to make my decision about how to respond very, very early in that encounter. This means that I will be more inclined to misread ambiguous signals. Second, the geographic proximity of those in state of nature encounters reduces response time, which further pressures my decision calculus. Finally, I have to presume that my interlocutor's command and control system is unstable in the sense that he is likely to be governed as much by his passions as reason. In that situation, I cannot assume that his signals mean what they appear to mean, or that he will not suddenly change his intentions, misreading perhaps one of my own signals.

A third risk is of rogue actors. There will be many in the state of nature who are sufficiently vainglorious that they will simply not respond rationally to whatever strategic game one is playing. Widespread superstition will only add to this problem. One can only assume that a reasonable percentage of those one encounters will be such rogues. Since no one profits by signaling that he or she is weak, the amount of time and the type of information available to distinguish between the rogue and the rational will be quite small. Any encounter must begin, therefore, with the background worry that one's interlocutor is not going to behave rationally.

Finally, any actor in the natural state will face a series of insoluble strategic dilemmas. Should I make defensive preparations? Developing defenses, such as a fortified compound, would introduce incentives for others to strike first in order to prevent my successful completion of those defenses. Still, if I could adequately defend myself, I could ensure adequate second-strike capabilities and would be at a considerable power advantage. However, not developing defenses would invite takeover by anyone who thought she was more powerful than me, including the vainglorious, who are mistaken on the point. A similar dilemma surrounds the acquisition of weapons. Gathering, building, or developing weapons sends the signal that I intend to use them, promoting defensive and noncooperative behavior in others. However, not developing weapons will make me appear conspicuously weak. Once anyone does obtain weapons, others will need to regain parity. The result might not exactly be an arms race, but the presence of more and more weapons would at least increase the risk of accidents and misperceptions. The only impossible move would be unilateral disarmament, as someone who does this "betrays himselfe to his enemy" (L 14.18, 96) in the worst possible example of irrational first performance. Finally, there is a dilemma about deception. If I do deceive and am caught, I confirm to others a weakness. If I do not deceive, I telegraph that weakness anyway. If I deceive and am not caught, I am showing off—and so attract the ambitions of the vainglorious.

Collective rationality, then, requires a sovereign, as such collective rationality is going to be very, very hard to achieve in the state of nature. Hobbes does not cite an actual historical example of the natural state because he does not think he needs to. On his own terms, the various factors influencing the decisions of those unfortunate enough to be in the natural state would be sufficient to see to it that they could never get out of it on their own. In this sense, it would be detrimental to Hobbes's point if he could show a historical moment when there was a generalized state of nature: such a moment would be a strong counterexample to be explained away. Hobbes's counterfactual runs the other way:

> If we could suppose a great Multitude of men to consent in the observation of Justice . . . without a common Power to keep them all in awe; we might as well suppose all Man-kind to do the same, and then there neither would be, nor need to be any Civill Government, or Commonwealth at all; because there would be Peace without subjection. (L 17.4, 118–19)

But of course people do have and require civil government. Hobbes invites those who do not trust his "inference, made from the Passions" to "have the same confirmed by Experience" (L 13.10, 89).[51] And, as Hobbes reminds us, that even in civil society we lock our doors and chests suggests that we are more sympathetic to his account than we want to admit.

Chapter 6

Constructing Politics

Those things which belong to Jurisdiction and Peace, and those things that are annexed to Justice, and Peace, appertain to none but to the Crown and Dignity of the King, nor can be separated from the Crown, nor be possest by a private Person.
Hobbes (1971, pp. 74–5)

The Hobbesian natural condition describes a condition of axiomatic disorder, stipulated in order to establish the conditions for successful government. As such an exercise in political philosophy, it is not intended to answer to any particular historical condition, although Hobbes is clearly pleased to draw his readers' attention to historical periods whose clear undesirability serves as persuasive machinery to underscore the desirability of the Leviathan as a solution. Indeed, as numerous commentators have noted, the desirability life in the Leviathan state is conditioned heavily by the undesirability of its alternative, the natural condition. The natural condition is overdetermined by the lack of a stable system of signification; many of the specific problems Hobbes identifies there, such as the inability of actors to contract, are traceable to that larger problem. Any transition out of the natural condition will therefore involve above all remedying the problem with signification. In this chapter, I want to look at two aspects of such a transition. The first is the importance Hobbes attaches to differentiating between the "multitude" and the "people." Consideration of this differentiation will underscore the extent to which Hobbes sets his own political thought against predecessors that based themselves on natural difference, rather than stipulative equality. The second aspect is the way that the law in general, and the judiciary in particular, function within a Hobbesian sovereignty. The judiciary's control of meaning is emblematic of the sort of tasks that the commonwealth must perform and serves an essential role in its stabilization in the quotidian lives of its subjects.

6.1 From Multitude to People

The signifier which identifies those in this condition of axiomatic disorder is "multitude [*multitudo*]," a term that bears the weight of describing the indescribable: people in their "natural condition." As I have suggested, the condition for the intelligibility of the natural state is, according to Hobbes, precisely its reduction to an undifferentiated disorder. His description of the state of nature thus serves the function of turning natural people into a homogeneous quantity, of the sort required for geometric calculation. On the one hand, then, the multitude names a discursive field which answers to the requirements of geometry. On Hobbesian grounds, because geometry is tied to quantification and quantification to counting, it is necessary to specify that all of those which are to be counted are of the same type. On the other hand, the specification of the multitude answers to the requirements of Hobbesian nominalism. Not only does it image people in a condition where meanings are unstable and then derive the consequences of what Hobbes takes to be a fact about language, showing in the process that the control of signification is a necessary condition for civil order, but also it allows the imposition of new names and a new system of political thinking without the baggage of historically sedimented meanings. Of course, it is one thing to erase historical meanings in the theoretical space of political philosophy and quite another to eliminate that baggage in the quotidian administration of the commonwealth. However, the strategy for the former begets a strategy for the latter, and the Hobbesian commonwealth is concerned to ensure the continual control of the processes of signification. I will return to this point in the final section of the chapter, detailing the daily operation of the judicature. In the remainder of this section, I would like to consider the contextual resonances of Hobbes's usage of "multitude" and how he marshals those in transforming the multitude into a "*populus*."

Hobbes's split between the multitude and people, which tracks the split between the state of nature and social contract, or a political philosophy founded on natural differences and one committed to ignoring them, is self-consciously innovatory. The split makes its first detailed appearance in *De Cive*, where he complains that political theorists "do not make a clear enough distinction between a *people* and a *multitude*" (DC XII.8). He continues:

> A *people* is a *single* entity, with a *single will*; you can attribute *an act* to it. None of this can be said of a multitude. In every commonwealth the *People* reigns; for even in *Monarchies* the *People* exercises power [*imperat*];

> for the *people* wills through the will of *one man*. But the citizens, i.e. the subjects, are a *multitude*. In a *Democracy* and in an *Aristocracy* the citizens are the *multitude*, but the *council* is the *people*; in a *Monarchy* the subjects are the *multitude*, and (paradoxically) the *King* is the *people*. Ordinary people and others who do not notice this point, always speak of a *large number* of men as the *people*, i.e. as the *commonwealth*; they speak of the *commonwealth* having rebelled against the *king* (which is impossible) and of the *people* wanting, or not wanting, what malcontent and murmuring subjects want or do not want; under this label of the *people*, they are setting the *citizens* against the *commonwealth*, i.e. the *Multitude* against the *people*. (DC XII.8)

As one might expect from the concept that applies to those in the state of nature, the multitude is defined by its lack of unity. Hence the multitude is insufficiently coherent to be the grammatical subject of a verb: only a people can act. In this sense, the only predication possible of a multitude is negative.[1] The contrasting "people" is a juridical fiction whose agency can be located in any number of actually existing entities; it functions as the formal cause which turns those actually existing entities into actors. The principle of civic organization therefore cannot be aggregation, and it is a mistake to measure a people as the aggregation of individual subjects; rather, it is to be understood according to its unity, as something definite as opposed to the indefinite multitude.

The distinction also appears in the earlier *Elements of Law*, where Hobbes again notes that "the controversies that arise concerning the right of the people, proceed from the equivocation of the word." He distinguishes two significations. The first "signifieth only a number of men, distinguished by the place of their habitation . . . which is no more, but the multitude of those particular persons that inhabit those regions, without consideration of any contracts or covenants amongst them." The other sense, as in *Leviathan*, is a political one, and "signifieth a person civil, that is to say, either one man, or one council, in the will whereof is included and involved the will of every one in particular." Hence, if one dissolves the house of parliament, it ceases to be the "people," becoming "the aggregate, or multitude of the particular men there sitting." The notion of meaningless aggregation recurs later in the paragraph, where Hobbes avers that "the heap, or multitude, cannot be said to demand or have a right to any thing." He thus concludes that "for where every man hath his right distinct, there is nothing left for the multitude to have right unto" (EL 21.11).[2] The passage leads one in similar interpretive directions as the one in *De Cive*; here, one may also note

Hobbes's deployment of terms of quantity. On the side of the people, one finds measurable particularity; on the side of the multitude, only various forms of aggregation—but not of successful aggregation, only the recognition of an aggregate, a "heap."

Hobbes is working within seventeenth-century conventions here, and his use seems to radicalize a thought in Grotius, who suggests that "in a like manner the Greeks derived their word for 'war' from a word meaning 'multitude'" (1962, II.I.ii.2). More generally, blaming the "mob" for disorder was a commonplace, particularly among Tacitist writers.[3] After Hobbes, and perhaps partly because of Hobbes's distinction between the crowd and the people, Whig writers expended considerable effort to prove that the "people" to which they referred had nothing to do with the rabble denigrated by royalist writers. The importance of Hobbes's distinction can be seen in a remark of Ireton's at the Putney debates, where he accuses the Levellers of insisting on natural equality: I am "afraid and do tremble at the boundless and endless consequences" of invocation of natural equality, since such invocation will devolve into "that wild or vast notion of what in every mans' conception is just or unjust."[4] In other words, having jettisoned any sort of Aristotelian notion of natural status, it is necessary to have a principle of order. Hobbes's gambit is that the disorder of the multitude will justify an authoritarian state. Later Royalists accused the Whigs of basing their political authority on "appeals . . . to the rabble" and the "ignorant and confused multitude" and "the inconsiderate rabble," or simply, "the multitude."[5] Locke's ambivalence on the subject is perhaps best noted by indicating the difficulty in deciding the extent to which he would in fact endorse a principle of equality: should the *Two Treatises* be read as a revolutionary manifesto demanding, *inter alia*, universal suffrage, or does it subtly reinscribe patriarchal rule?[6]

In any event, Hobbes is quite clear on the point. Natural differences cannot be the basis of political theory. Because these natural differences are not to be taken into account, civil society can only function when a principle of order is imposed on the multitude. The basic problem is one of attribution: how can one know that the actions of a crowd should be taken as representing the actions of its members? One would need some sort of decision rule for such an attribution, but since the same set of activities might be legitimately taken as the group's action in one setting but not in another, it is impossible to know what the group's action is, absent an explicitly articulated decision rule. For example, a majority vote might count as a decision in some groups (or for some decisions of a single group), but different groups or different decisions by the same group might require a supermajority or be subject to a veto by certain members. The depth of the problem becomes clear when one considers unanimous actions by a crowd, for example the storming of the Bastille. Consider a prosaic example first. It is one thing when everyone gets off the bus at the same time due to the clear variety of reasons they do so. Such a mass departure from the bus is the aggregated actions of several people, not a group action. Surely Hobbes does not mean to say that unanimous actions for the sake of the group do not count as actions of the group, such that, at least here, the multitude can act?[7] He clearly does, however. How does one know why the members of the mob are storming the Bastille, and do all of the members even intend to storm the Bastille? What is the intention behind storming the Bastille—to free prisoners, to make a political statement?

The first thing to notice is that the epistemological problem is exactly the one Hobbes identifies with analysis in general, with movements from particular phenomena to the rules of their generation: how does one know that one has selected the correct rule? In the case of a political theory based on natural hierarchy, one derives the decision rule from the hierarchy. With that decision rule removed, it has to be instated artificially. Hence, in *Leviathan*'s introduction, Hobbes indicated that "the *Pacts* and *Covenants*, by which the parts of this Body Politique were at first made, set together, and united, resemble that *Fiat*, or the *Let us make man*, pronounced by God in the Creation" (L Intro., 9–10). The analogy is not pointing to the fact that God is powerful, so much as to the fact that God's creation was *ex nihilo*: there was nothing called "man" prior to God's fiat, and the power of the creation is precisely in that the phenomenon of man is brought forth from something that does not resemble it in any way.

Conceptual ballast for understanding these uses of "multitude" is also available from other areas of Hobbes's writings. First, "multitude" has Biblical connotations, having to do with excessive, meaningless speech. In *Job* 11:2, Zophar wonders, according to the King James edition, "should not the multitude of words be answered?" The answer, of course, is that they should not: not, as Zophar thinks, because Job must have sinned, but because demanding an answer of God is itself a sin. The Vulgate offers an interpretive clue: "*numquid qui multa loquitur non et audiet aut vir verbosus iustificabitur*"—we are to inquire whether someone who speaks many things but does not hear, that is, whether a person "full of words"—is justified. Job's problem is that his own speech does not justify itself, and there is difficulty in going from the fact that he is afflicted to the reason why. The problem is thus one of relating principle to an observed aggregation of things, and of deciding whether the things themselves can supply the principle. The usage of "multitude" to translate this passage was standard. In the words of one popular seventeenth-century commentator, "a multitude of words is sinful" when they are "unprofitable, light, vain, frothy, words that have no nourishment in them" (Caryl, 1677, I, 1017). Hobbes, of course, picks up on the Job reference and gestures frequently to the problem of such speech in *Leviathan*.

Second, Hobbes also uses "multitude" in his physics to refer to the failure of measurability. For example, in his early *De Motu*, he writes of uses of "infinite" and refers to the "possibility of infinity (in multitude) [*infiniti (multitudine) possibilitatem*]" (DM 19.6). As in the case of Job's words, the key point here is again the mismatch between the aggregation and the ability to apply principle to it. An infinite quantity is precisely that which, according to Hobbes, presents an impossibility of understanding. Hence, in *Leviathan*, he remarks that we cannot form an image of an infinite magnitude, and the term is used when "we are not able to conceive the ends, and bounds of the thing named" (L 3.12, 23). How to move from an aggregation to its principle of intelligibility was of course the regressus problem, and it should be noted here that terms of multitude and profusion were often applied to the generation of principles prior to their reduction into a demonstration *propter quid*, as for example when Zabarella says of an early step in the process that it "treats the way whereby we abound in demonstrations from the same places of cognition" (*De Regressu*, 1222, referring to Aristotle PA 2.14).

Leaving the multitude behind, then, constructs the grammar of political philosophy. This *poietic* principle has the effect that, for the Hobbesian apparatus, since there is no *arche* given in the posited state of nature, the created political person, who can be said to have been brought forth from the state of nature, can be purely governed according to the political *arche*. There will be no extrapolitical archic ghosts left to tag along. All of the problems of historical memory which plague, for example, Machiavelli, disappear by fiat. They might remain as problems of day to day governance, but they disappear as theoretical problems. They become, instead, practical matters to be addressed by institutions.

The greatest risk to the sovereignty is *doxa*, which will achieve almost demonic status. Thus, a political theory which proceeds by means of commentaries becomes unable to account for its own origin and reveals itself as meaningless: "Commentaries are commonly more subject to cavill, than the Text; and therefore need other Commentaries; and so there will be no end of such Interpretation" (L 26.25, 193).[8] It is significant that this passage occurs in a discussion of the interpretation of "written laws." We will return to the issue of legal interpretation shortly, but for now it should be noted that a table of definitions functions as axiomatic to such a "written law." By codifying the meaning of words, Hobbes removes the ambiguities of understanding which authorize textual commentaries, and in so doing, ensures that the space of codification remains a regulated one. Hence, when he follows that "the significations of almost all words, are either in themselves, or in the metaphoricall use of them, ambiguous; and may be drawn in argument, to make many senses; but there is onely one sense of the Law" (L 26.26, 194), we should also read this as a methodological claim about a science itself: in order to avoid a proliferation of ambiguities, it is to be well-governed.

The need to avoid ambiguous meanings then, becomes a recurrent theme in the Hobbesian political apparatus, and the possibility of impure and "metaphoric" meanings will haunt it incessantly. The risk is always the collapse of the discourse *qua* science into an instability of meaning which legitimates the intrusion of an endless series of quasi-meaningful ghost-signifiers, whose hauntings render the discourse radically unstable. Such unstable discourse is immediately accused of insanity. Thus, of a text of Suárez's, Hobbes remarks: "when men write whole volumes of such stuffe, are they not Mad, or intend to make others so?" (L 8.27, 59). The insanity occurs when word meanings are not well grounded, when the table collapses, when judgment therefore gives way to fancy: "without Steadinesse, and Direction to some End, a great Fancy is one kind of Madnesse" (L 8.3, 51). Commentaries, or a discourse which encourages them, risk destabilizing the entire apparatus.

Hobbes is thus able to say that the usage of "senselesse and ambiguous words" has political consequences: "reasoning upon them, is wandering amongst innumerable absurdities; and their end, contention, and sedition, or contempt" (L 5.20, 36).[9] *Behemoth* reads like a catalogue listing of those who have traded upon ambiguous words to promote their own seditious plans. In particular, the universities are to be reformed to teach *Leviathan* and not scholastic commentaries; at the end of that text, he suggests that "it may be . . . profitably taught in the Universities . . . seeing the Universities are the fountains of Civill, and Moral Doctrine" (L R&C 16, 491). By *Behemoth*, the failure of the universities in this regard has earned them excoriation:

> It is a hard matter for men, who do all think highly of their own wits, when they have also acquired the learning of the university, to be persuaded that they want any ability requisite for the government of a commonwealth, especially having read the glorious histories and the sententious

> politics of the ancient popular governments of the Greeks and Romans, amongst whom kings were hated and branded with the name of tyrants, and popular government (though no tyrant was ever so cruel as a popular assembly) passed by the name of liberty. (1990, p. 23)

Again, the misuse of names is central: the name of liberty has been wrongly attributed by the Greeks and Romans to living under a popular government, and belief in that has *caused* the institution of the worst kinds of governance imaginable. It is perhaps in this light that we should read Hobbes's remark that "there was never any thing so deerly bought, as these Western parts have bought the learning of the Greek and Latine tongues" (L 21.9, 150).[10]

Things thus turn on what Deleuze calls a "Platonism" of the seventeenth century.[11] Words proliferate and are always copies of some sort of other meaning; the fiction of universal language such as Biblical Hebrew is untenable, and the psychology involving an intellectual faculty has succumbed to Ockham's razor. The trick, therefore, is to distinguish the good kinds of copies from the pretenders. In particular, previous meanings of words are to be reconfigured, since their outright banishment is impossible. Without proper definitions, that is, without establishing the archic matrix through which all judgment occurs, we will be subject to "fancies" and ghosts, have no way of knowing if we are asleep or awake, and have no way of knowing whether anything exists. In short, there would be no phenomena at all. We would induce the equivalent of a state of nature in ourselves, from which we could not emerge without the help of a sovereign to provide a conceptual structure for us. If there are to be phenomena, if *to fainesthai* is to present itself, this can only occur through a conceptual apparatus secured by definitions. Individuals cannot function in an originary state, but the sociality of language means that accurate definitions can divert one from the slippery slope into such confusion.

The conceptual apparatus is therefore to be hermetic: the more tightly sealed it can be against the outside, the less the risk of invasion by such seditious words. It is perhaps in this regard that we can begin to understand why Hobbesian punishments, particularly against things labeled "seditious" are so harsh. They concern the possibility of politics itself. I will return in a moment to the question of judging, but let me note that even here the importance lies in the manner by which the sovereign is able to provide its own conceptual apparatus, *internal to its definition as sovereign*, by which to ensure this hermeticity. In this regard, the positing of a state of nature becomes the only means by which the origin of political philosophy can be made secure. If a commonwealth were to adopt the model of the *Leviathan*, and the citizens were to learn its doctrines, then the memory of past liberties would themselves become unknowable; the ladder by which the sovereign becomes established as such would erase itself, leaving only the sovereign as having "always been there." Furthermore, the interposition of the dichotomy between a meaningless state of nature and a precisely regulated political space into a conceptual apparatus which has begun by establishing the importance of definitions subtly transfers all other political philosophies to the state of nature, where it is better to reduce whatever differences they might actually have as if they were all the same.

6.2 Judging Judges

Hobbes's understanding of the commonwealth as essentially a place of regulated meanings designed to overcome the infelicities in the state of nature has implications for day-to-day life in the Leviathan state. Examination of these implications is important for several reasons. First, it indicates why Hobbes spends so much of his time talking about institutions within the sovereignty, in particular those which are generative of meaning: universities, churches, and the judiciary. In what follows, I will focus on the judiciary because it is here that the Hobbesian sovereignty most clearly makes meaning in the lives of its subjects and because the functioning of the judiciary is exemplary of how the Hobbesian apparatus ensures its preservation by replicating itself in specific contexts.[12] A preliminary sketch of the importance of the judiciary to Hobbes follows from his understanding of law. The civil law, he says, "is to every subject, those rules which the commonwealth hath commanded him . . . to make use of, for the distinction of right and wrong, that is to say, of what is contrary, and what is not contrary, to the rule" (L 26.3, 183, italics omitted).[13] Law, however, is text and so requires interpretation: "all laws, written and unwritten, have need of interpretation" (L 26.21, 190). Since the multiplicity of private centers of interpretation of right and wrong is one of the principle defects of the state of nature, the commonwealth will require authorized public agents to carry out this function. These agents carry out acts of judging, which Hobbes defines as follows: "*judging* is nothing other than applying *laws* to single cases by *interpretation* [*Iudicare* enim nihil aliud est quam *leges* singulis casibus *interpretando* applicare]" (DC XIV.13).

Second, as Hobbes's definition of the judicial act indicates, standard questions of "theory" and "practice" are understood by Hobbes fundamentally to be questions of interpretation.[14] There is, in other words, no mechanical application of the law to individual cases. Rather, such interpretation will always be governed by the same principles of reason which were employed in the construction of the sovereignty in the first place, and these principles entail (as I will argue) the usage of a thoroughly Aristotelian sense of equity as a corrective to mechanical application. Again, a preliminary sense of the importance of the subordination of theory and practice to hermeneutics is evident from a passage early in *Leviathan*'s chapter on civil laws. There, Hobbes takes it as a point of theoretical consensus among his contemporaries that "laws can never be against reason" and that "not the letter (that is, every construction of it) but that which is according to the intent of the legislator, is the law." The issue, he suggests, is "of whose reason it is that shall be received for law" (L 26.11, 186).[15] The complexity of this issue derives from Hobbes's understanding of personation: although "in all courts of justice, the sovereign (which is the person of the commonwealth) is he that judgeth" (*ibid.*), the person of the commonwealth is an artificial person, and may or may not be coextensive with any natural person. As Hobbes adds emphatically, "it is manifest that the authority of the law does not depend on the private reason of a person" (OL III, 199; cf. L 26.11, 187 for a slightly less emphatic statement).

Third, and as a consequence, it allows one to reframe concerns about Hobbesian absolutism. Let us be clear: Hobbes advocates an *absolute* state. However, this state is not to be *arbitrary*, and the point of the institutions of sovereignty is to ensure the complete integration of subjects into the commonwealth, not to allow unregulated violence against them. In principle, then, the Hobbesian sovereign, and thus the sovereignty itself, is limited to powers which can be derived from equity and the *salus populi*. Although Hobbes is incapable of guaranteeing that the sovereign does not become corrupt, the degree to which the sovereignty is instantiated in the quotidian life of institutions creates a substantial bureaucratic check against such imperial ambitions. In slightly different terms, the (natural person of the) sovereign himself is a part of the Hobbesian sovereignty, not the sovereignty itself.[16]

To approach the question of judicial interpretation is thus to approach the question of limits to the Hobbesian sovereign. The caricature of Hobbes as authorizing an arbitrary and absolute sovereign has a venerable tradition.[17] It is, however, important to distinguish between the sense in which the sovereign is absolute and the sense in which it is arbitrary. The importance of this distinction emerges quite clearly in Rousseau who, after complaining about authoritarian tendencies in Hobbes, declares that when someone's individual will is at variance with the general will, the individual will must be bent to the general, and that the person in question must thereby "be forced to be free" (*Social Contract* 1.7). For his part, Kant (who also complains about authoritarianism in Hobbes) opines that one should establish "a society in which freedom under external laws would be combined to the greatest possible extent with irresistible force" because "man is an animal who needs a master" (1991, V–VI, pp. 45–6).[18] Neither Kant nor Rousseau found any inconsistency in forcing people to be free, but both thought that Hobbes gave free license to despotism.

A more nuanced version of the caricature can be found in Locke. In arguing that, in the state of nature, everyone has the right to punish transgressions of the natural law, Locke considers the objection that

> It is unreasonable for men to be judges in their own cases, that self-love will make men partial to themselves and their friends, and that on the other side, that ill-nature, passion, and revenge will carry them too far in punishing others, and hence nothing but confusion and disorder will follow, and that therefore God has certainly appointed government to restrain the partiality and violence of men. (2T, 13)

Although Hobbes argues that right reason, rather than divine command, should convince someone of the desirability of government, the passage seems clearly also to be directed at Hobbes, since it is precisely the right to private judgment that Hobbes says cannot exist in civil society. Since Locke is concerned to defend the proposition that the state of nature is not *absolutely* bad, that (for example) intelligible regimes of property can exist in the state of nature understood as a commons, the objection here might be rephrased as follows: would not assigning to everyone in the state of nature the right to enforce the laws of nature guarantee that the state of nature be a Hobbesian *bellum omnium contra omnes*? If so, the state of nature would be unable to do the conceptual work Locke requires of it.

In response, Locke immediately concedes that civil government is the "proper remedy for the inconveniences of the state of nature." He then turns the argument into a condemnation of absolute monarchy (2T, 13; repeated at 2T, 91): if it is the case that private judgment is bad, one should remember that "absolute monarchs are but men," and would therefore exercise their own private judgment. Such a situation, in which "one man commanding a multitude has the liberty to be judge in his own case, and may do to all his subjects whatever he pleases," would for two reasons be worse than the state of nature. First, at least in the state of nature, one is not bound to follow the arbitrary dictates of private judgment. Second, in the state of nature, someone who makes a bad judgment is "answerable for it to the rest of mankind." Hence, absolute monarchy takes all the disadvantages to the exercise of private judgment which would exist in the state of nature, and then removes all the checks and balances to those disadvantages. One strategy to avoid this problem—and the strategy which I will argue that Hobbes follows—is to institutionalize the performance of the sovereignty. To say that the absolute sovereign is a "private" person is to say in Hobbesian terms that the natural person of the monarch is at the same time the person of the commonwealth. Hobbes rejects precisely this conflation of personae. That is, although the sovereign is absolute, its actions are limited by constraints which are written into its definition.

To see these constraints, consider the way that Hobbes reworks the traditional uses of proportion in the definition of justice. In his discussion of distributive and commutative justice, he discusses what he takes to be the standard division of the two: commutative, an arithmetic proportion, is "in the equality of value of the things contracted for;" whereas distributive, a geometric proportion, is "in the distribution of equall benefit, to men of equall merit." To the first, he objects that the value needs to be conceived in terms of what the parties to a contract are "contented to give," and that that commutative justice in general is thus the "Justice of a Contractor" (L 15.14, 105).[19] Distributive justice, he says, is the "justice of an Arbitrator; that is to say, the act of defining what is Just." He follows that if the arbitrator performs properly, "he is said to distribute to every man his own" and that this performance is better called "equity" than distributive justice (L 15.15, 105).

In other words, that arbitration is a function of the sovereignty puts the sovereign necessarily in the role of providing distributive justice, and this distributive justice is a proportion governed by the rules of equity. Equity entails, *inter alia*, impartial judgment, and it is the eleventh law of nature that "if a man be trusted to judge between man and man, he should deale Equally between them" (L 15.23, 108). Without this equity, the entire "use of Judges, and Arbitrators" (*ibid.*) is undermined, as a bad judge deters others from seeking judges and arbitration; this then pushes them to violent techniques for conflict resolution. All of these precepts are clearly foundational to the entire Hobbesian enterprise: the tenth through nineteenth laws of nature, which provide rules for arbitration (and the sixteenth, of submission to arbitration), are all said to follow from the ninth. The ninth, as I have indicated, is that people recognize each other as equals, and is the one Hobbes directs specifically against Aristotle.[20] In other words, the gesture by which Hobbes overthrows the Aristotelian "slaves by nature" line also establishes an initial defense against the charge that the Leviathan state is an arbitrary dictatorship. To be within the Hobbesian system is already to have committed to the rejection of the sort of arbitrary rule about which Locke is worried. For Hobbes, arbitrary dictatorship is an artifact of Aristotelianism, not a consequence of his own system. This is perhaps why, unlike a plurality of his ancient interlocutors, Hobbes has no sustained treatment of tyranny or tyrants, except to reduce "tyrant" to a pejorative used by those who are not content with their rulers.[21]

Once one views people as equal, and as seeking their own preservation, then one has to suppose that they intend to be in a livable commonwealth. Hence, as a corollary of equity, Hobbes declares that no member of the commonwealth is to arrogate rights that he would not allow others to have:

> As it is necessary for all men that seek peace, to lay down certaine Rights of Nature; that is to say, not to have libertie to do all they list: so it is necessarie for mans life, to retaine some; as right to governe their owne bodies; enjoy aire, water, motion, waies to go from place to place; and all things else, withouth which a man cannot live, or cannot live well. (L 15.22, 107)

It is of course one thing to design a commonwealth with this in mind and another thing to achieve it in result. It is nonetheless worth noting that this is Hobbes's design. The implementation of that design is left to the various institutions of the sovereignty. Since the question here is one of equity and arbitration, it is necessary to consider the office of the sovereignty which performs the function of securing equity and arbitration, the judiciary.

The seventh right of sovereigns by institution is the "Right of Judicature," defined as "of hearing and deciding all Controversies, which may arise concerning Law, either Civill, or Naturall, or concerning Fact" (L 18.11, 125). The sovereign is granted this right by a reductio of its denial: if there were no decision of controversies, there would be no protection of subjects against the injuries to one another, and therefore everyone would retain the "right of protecting himselfe by private strength, which is the condition of Warre; and contrary to the end for which every Common-wealth is instituted" (*ibid.*).[22] Insofar as judges are to apply the law in specific situations, their role is interpretive, and they exist at the intersection of the sovereign considered *qua* sovereign and the sovereign considered *qua* governor of specific actions. For Hobbes, the distinction between the two is one of reason alone: the point of the sovereign being able to authorize its own utterances and to seal off other competing utterances is that to be a sovereign is precisely to be the governor of specific actions. The interference of competing discourses—historical memory, university professors, Latin orators, and the like—is eliminated by the sovereign's overdetermination of the political space; this in turn eliminates the gap between being the sovereign in general and the sovereign in particular instances. In this way, on the one hand, the transformation of the multitude into a people is effected as individuals are brought into the space of the sovereign authority and out of the state of nature. On the other hand, the judges appear not only as interpreters of the sovereign but also as specific instantiations of it, and the archic role of the judges will be to repeat and re-present the sovereign. In this representation, the judges will bring forward the words of the law in all of their clarity and according to their origin, in the geometric terms of the sovereignty. Hobbes writes, in a passage we have already considered in part:

> The significations of almost all words, are either in themselves, or in the metaphoricall use of them, ambiguous; and may be drawn in argument, to make many senses; but there is only one sense of the Law . . . the literall sense is that, which the Legislator intended, should by the letter of the Law be signified. Now the Intention of the Legislator is alwayes supposed to be Equity: For it were a great contumely for a Judge to think otherwise of the Soveraigne. He ought therefore, if the Words of the Law do not fully authorize a reasonable Sentence, to supply it with the Law of Nature; or if the case be difficult, to respit Judgement till he have received more ample authority. (L 26.26, 194)

The passage is difficult, and the later Latin text clarifies matters somewhat. There, Hobbes writes:

> The letter or word of the law taken *per se*, they call the literal sense; but they say the sentence of the law is what the legislator would have wanted to be understood. And indeed they are correctly distinguished. But what would the legislator have wanted understood through the words of the law, [and] from where is this understood? It is understood from what the legislator is, i.e., the person of the commonwealth, who is always to be understood to want that which is equal. And, if the words themselves do not suggest an equal sentence, the laws of nature are to be consulted. That someone is not able to have remedy against an injury is to be supposed to be against the will of the legislator. (OL III, 204–5)

Hobbes thus outlines a procedure for interpretation, one that attempts to strike a middle ground between an excessive literalism and the slippage into the endless interpretive battles that mark a common law regime.

The guard against both of these extremes is "equity." Hobbes's use of this term functions in a manner that both echoes and transforms its sense in Plato's *Statesman* and Aristotle's *Ethics*, as a "correction of what is legally just" (*Ethics* 1137b12). For Aristotle, equity allows a judge to remedy a difficulty in the application of the law: it must speak in universal terms, but be applied to particular cases, which may not fit the universal terms. "Equity" thus stands for what the lawgiver would have decided about a particular case, had he known its details, and generally serves as a remedy against the text of the law for the sake of justice. Grotius invokes this sense of the term in arguing that natural reason is useful in judging intent, "since all contingencies neither be seen nor set forth, a degree of freedom is needed in order to make exceptions of cases which the person who has spoken would make an exception of, if he were present," though he urges caution in invoking this principle (1962, II.XVI.xvi.1). Hobbes clearly refers to this meaning of "equity," and adds to it a sense in which equity provides an imperative for the interpretation of the law itself and not just its application to particular cases. Thus, if the words of the law themselves do not suggest equity, right reason, about which Hobbes has many fewer reservations than Grotius, and which, one should recall, is always in favor of equity—is to be consulted. The final sentence or judgment of the law is a product both of finding out what the words of the law say and of applying the standards of right reason to them.

Trying to get the words to speak their own meaning, to write commentaries on them, is an invitation to disaster, since "commentaries are commonly more subject to cavill, than the Text; and therefore need other Commentaries; and so there will be no end of such Interpretation" (L 26.25, 193). "End" should be understood in two ways: first, in the sense that there will be no conclusion to the process of writing commentaries; and, second, that there will therefore be no purpose to the activity. As I have indicated, it is pointless, Hobbes thinks, to study examples and texts without knowing what principle governs them. This is a particular problem for the common law, since its principles are to be derived precisely from a series of example cases. In other words, it is no surprise that Hobbes would be opposed to the common law. Hobbes was also not alone in using equity as a check against proliferate legal interpretations. In his discussion of property, for example, Grotius defends the right to use another's property in the case of necessity. To defend this right, he says, we must first "consider what the intention was of those who first introduced individual ownership; and we are forced to believe that it was their intention to depart as little as possible from natural equity." This is the Aristotelian sense of equity; in the following sentence,

Grotius explicitly makes it a matter of statutory interpretation: "for as in this sense even written laws are to be interpreted, much more should such a point of view prevail in the interpretation of usages, which are not held to exact statement by the limitations of a written form" (1962, II.ii.vi.1). Grotius has a lengthy, example-based discussion of interpretation, and some of the principles he derives (such as the one to favor a narrow, rather than a wide, reading of the law) evidence a concern with the interpretive problems Hobbes addresses. Grotius's example of a licensed broad interpretation, outside the letter of a contract, specifically involves questions of intent. For example, an agreement that "a certain place shall not be surrounded with walls . . . made at a time when there was no other kind of fortification" implies that "it will not be permissible to surround that place even with an earthwork, if it is fully established that the sole reason why walls were prohibited lay in the intent that the place should not be fortified" (1962, II.xvi.xx.3). He then cites Cicero on the point:

> What then? Had sufficient provision been made for this in the words? Not at all. What is it then that is valid? The intent; and if this could not be understood by us without a speech we should not use words at all. Because it cannot be so understood, words were invented, not to hinder, but to express the intent. (1962, II.xvi.xx.4; citing *For Caecina* xviii.53, 59, 63)

It is this constellation of issues that Hobbes addresses.

Two further keys to the *Leviathan* passage emerge from the scholastic terms it incorporates. A "sentence [*sententia*]" is a statement or judgment; in scholastic commentaries, *sententiae* were often invoked to help one deal with specific questions. To "supposit," in the tradition developed in thinkers such as Ockham, is to "stand for," in the sense that a mental act stands for something *extra animam*.[23] In Hobbes's case, supposition is governed by the grammar of the sovereignty. This claim has both a stronger and a weaker sense. In the weaker sense, for a judge to think that a sovereign did *not* intend equity is to disrespect the sovereign; judges should always respect the sovereign; therefore equity is to supposit for the sovereign intent. The ascription of equity is thus part of what it means to judge well. In the stronger sense, judges are themselves representatives of the artificial person which is the state and are thus aspects of the sovereignty. As with any other proxy, artificial persons are bound by the rules of their commission to act in certain prescribed ways; when duly authorized judges thereby apply the law, the commonwealth can be said to have authored the sentence of the law.[24]

The rule for supposition thus solves what would otherwise be the Platonic problem of interpretation, of knowing that one's reading of a text comported with what its author intended. Hobbes specifically collapses the spoken/written distinction: any successful sign of the legislator as to what is to be done suffices in this regard, and "non-written" law is reserved for natural law and the dictates of reason (DC XIV.14). In place of the Platonic derogation of texts and privileging of either conversation with the author or intuitive contemplation of an *eidos*, Hobbes substitutes a constitutive rule of equity. Because this rule is constitutive and internal to the sovereign apparatus, Hobbes's substitution also solves the skeptical/Machiavellian problem of knowing that one's representation of the external *eidos* is veridical. Words and things align perfectly in the space of the "law," and that means that there is no possibility of ambiguity in its signification, when correctly applied by the judiciary.

When the intent of the legislator (sovereign) is always to be equity, and it were a "great contumely" to think otherwise, Hobbes is pointing to the manner in which the sovereign operates as the representative of the "people," which are themselves constituted as such by the original contract. The sovereign is created by the act of leveling out specific differences for the sake of creating a political space, and in that political space, the sovereign is the representation of all of its subjects' political existence. The people as such are distinct but not different from one another:

> The only way to erect such a Common Power . . . is, to conferre all their power and strength upon one Man, or upon one Assembly of men, that may reduce all their Wills, by plurality of voices, unto one Will . . . This is more than Consent, or Concord; it is a reall Unitie of them all, in one and the same Person, made by Covenant of every man with every man, in such manner, as if every man should say to every man, *I Authorise and give up my Right of Governing my selfe, to this Man, or to this Assembly of men, on this condition, that thou give up thy Right to him, and Authorise all his Actions in like manner.* (L 17.13, 120)

In other words, the only way to establish (and maintain) a commonwealth in which life is no longer "nasty brutish and short" is to transfer the posited equality of people in the state of nature to a single representative, which is the artificial person of the state. In this way, their equality is maintained: all have equal power (or lack thereof) in relation to each other, and all are equally powerless before whoever represents the sovereignty. Because the power is united into a single, univocal agent, the tendency of people's individual wills to degenerate into war is made definitionally impossible.

The sovereign will therefore institute law, defined as "the word of him, that by right hath command over others" (L 15.41, 111).[25] Since the meaning of laws carries necessary ambiguity, the sovereign will also appoint authorized interpreters, and "when there is a question about the *meaning of laws* [*sententia legum*], it is to be sought from those whom the sovereign has charged with the *trial of cases,* the *courts*" (DC XIV.13). As acting on commission from the sovereign, who is in turn the representative of the people, the judges are to act in terms dictated by the original structure of the commonwealth, which is the effort (of those who are willing to view each other as equals) to secure their own preservation. Thus, "the finall Cause, End, or Designe of men . . . (in which wee see them live in Commonwealths,) is the foresight of their own preservation, and a more contented life thereby" (L 17.1, 117). Although the sovereign is not subject to the positive laws it promulgates, the "law" of the sovereignty is thus *salus populi*; as Hobbes puts it in *De Cive,* "all of the duties of sovereigns are implicit in this one phrase: *the safety of the people is the supreme law*" (DC XIII.2).

In this sense, the law functions as the rule or measure which transforms the unintelligible actions of the members of the multitude to the actions of the people. Hence,

> It is the responsibility of the same *Sovereign power* to come up with rules or measures that will be common to all, and to publish them openly [*communes omnibus exhibere regulas, siue mensuras, & publicè eas declarare*], so that each man may know by them what he should call [*appellandum sit*] his *own* and what *another*'s, what he should call *just* and *unjust, honourable* and *dishonourable,* good and bad; in summary, what he should *do* and what he should *avoid doing* in social life. (DC VI.9)

The sense of common rule or measure was a commonplace in scholasticism; as Thomas Aquinas puts it, "law is a certain measure and rule of acts whereby man is induced to act or is restrained from acting."[26] In Hobbes's case, the dictum is adapted to specify the nature of the transformation of the warring multitude into the peaceful people; the mechanism of this transformation is the introduction of order, in all the senses of the term. Most important, the order is one of signification: we are to discover rules for what ought to be called mine and yours. What we are not to do is to search for metaphysical or theological evidence of such property claims.[27]

Equity therefore assumes *archic* status for judges' decisions, at the explicit expense of custom or *stare decisis*: "because there is no Judge . . . but may erre in a Judgement of Equity; if afterward in another like case he find it more consonant to Equity to give a contrary Sentence, he is obliged to doe it" (L 26.24, 192).[28] The judge's behavior is to be regulated according to the conceptual apparatus of the sovereign and not anything else; Hobbes continues that "no mans error becomes his own Law; nor obliges him to persist in it. Neither . . . becomes it a Law to other Judges, though sworn to follow it" (L 26.24, 192). So too, if it is true that "all Laws, written, and unwritten, have need of Interpretation" (L 26.21, 190), one should not forget that "not the Letter . . . but that which is according to the Intention of the Legislator, is the law" (L 26.11, 186–7), because it would be the contumely of challenging the sovereign to think otherwise. The circle thus closes itself, and the possibility of interpretation according to the demons of historical custom disappears: the sovereign embodies equity, the law is the expression of the sovereign, and so the law has to be governed by equity; the judge interprets the law, the judge's essential function is to uphold the sovereign, and so the interpretation has to be according to equity and the principles of distributive justice. Anything else is simply not politics, and in this way the original *poietic* act by which the political is created replicates itself in further ones. Then, politics has responded to the aims of a science: its knowledge and subjects are secure. Arbitrary, Machiavellian displays of power are radically impossible in a Hobbesian commonwealth. Power instead comes firmly to rest in the production of its own subjects according to the principles of a disciplined, geometric thought.

Chapter 7

Conclusion: From Erasing Nature to Producing the Multitude

The fable of the war of all against all is as idiotic as all fables of origins. But behind this feeble tale of death and salvation, something more serious makes itself felt, the declaration of the ultimate secret of any social order, the pure and simple equality of anyone and everyone: there is no natural principle of domination by one person over another. The social order ultimately rests on the equality that is also its ruination.

Jacques Rancière (1999, p. 79)

7.1 Hobbesian Biopolitics

We return to where we began: the sense that Hobbes is our contemporary, that he is everywhere, as evidenced particularly by the cultural resonance of his state of nature. We are now, however, in a better position to adumbrate some reasons for Hobbes's ubiquity, which can be framed as a series of three propositions: first, that ours is a biopolitical age; second, that Hobbes is the first major theorist of biopolitics; and finally, that our age is thereby profoundly Hobbesian. Let us interpret each of these in turn.

7.1.1 Ours is a biopolitical age

This thesis, as articulated by a range of thinkers influenced by some of Foucault's later writings, I take not to require extensive defense in the present context. The thought that power is primarily about the productive management of life, rather than asserting authority over subjects primarily at their death, is both presupposed and explicitly defended across a wide range of theoretical discourse. It forms the basis not just of much recent "continental" political philosophy, but forms the necessary substructure of even mainstream work in Chicago-school analyses of economics and law. On the continental side, Hardt and Negri (see 2000, 2004) have produced perhaps the most prominent version of the thesis, arguing that both our understanding of the functioning of power and the ways to resists it need to be rethought in biopolitical terms. Markers of this need include the expansion of capital relations beyond the working day ("complete subsumption" of society by capital) and the increasing social value attached to immaterial goods—ideas, ways of life, and so on. From the Chicago school, Cass Sunstein (1996), for example, proposes that law can and should operate efficiently by fostering desirable social norms on such topics as seat belt use and recycling and by discouraging undesirable ones like littering. More generally, the law and economics literature, whose hegemony in the legal academy perhaps cannot be overstated, views problems of regulation as problems of efficiently reducing transaction costs: of reducing the various impediments to individuals' negotiating their way to individually satisfactory allocations of property and other entitlements.[1] Few of these papers make any reference to Foucault, but it was Foucault who proposed that

> This, I believe, is the essential issue in the establishment of the art of government: introduction of economy into political practice. And if this is the case in the sixteenth century, it remains so in the eighteenth. . . . To govern a state will therefore mean to apply economy. To set up an economy at the level of the entire state, which means exercising towards its inhabitants, and the wealth and behavior of each and all, a form of surveillance and control as attentive as that of the head of a family over his household and his goods. (1991, p. 92)

In any case, it is hard to disagree with pronouncements such as Hardt and Negri's that, in the current political order, "economic production and political constitution tend increasingly to coincide" (2000, p. 41). It is also not hard to see the initial plausibility of a connection with Hobbes: the state of nature, recall, is presented as a systematic removal of the accoutrements of social and economic life, a state with "no place for Industry . . . no Arts; no Letters, No Society" (L 13.9, 89).

Before pursuing the connection with Hobbes in more detail, it is perhaps worth recalling three further, surface-level features of biopolitical life that (as I argued in Chapter 2) find early articulation in Hobbes. First, biopower takes the social order and the individuals who inhabit it as fundamentally malleable and constructible. At the same time, it valorizes productive activity at both the level of individuals and the social order. Individuals become the object of disciplinary techniques exemplified not just by Bentham's panoptic prison but also by rigorous school curricula. The population itself becomes an object of study, as "problems of birthrate, longevity, public health, housing and migration" (Foucault, 1978, p. 140) move to the fore. Biopower is, in other words, deeply *poietic*, increasingly rejecting the idea that nature imposes limits on human plasticity and giving moral approval to productivity. Second, biopower treats the boundaries between art and nature as fundamentally fluid. Donna Haraway's "Cyborg Manifesto" (1991) is a particularly sharp formulation of this point; as Haraway indicates, initially citing Foucault, we can expect to see several kinds of boundary breakdowns: between animal and human, between organism and machine, and between physical and nonphysical. Conservative complaints about biotechnological research, the increasing importance of networks and information technologies to the economy, and even work in neuroscience confirm these social erosions. Finally, biopower rejects any notion of fundamental natural difference. We are all equal; this proposition has been so thoroughly absorbed into our political thought that it seems hard to imagine what a defense of it might look like or why one might be necessary.[2] Thinking as construction, the erosion of art/nature boundaries, and the *ethos* of equality can all be traced to Hobbes.

7.1.2 Hobbes is a thinker of biopolitics

This proposition is perhaps best explored by contrasting it with another, initially plausible one: Hobbes is an exemplary theorist of juridical power. *Prima facie* evidence in support of this contrasting proposition is easy enough to find in Hobbes's endorsement of absolute sovereignty. Hence, one might argue, the outcome of Hobbes's reformulation of civil philosophy according to the principles of construction turns out to be familiar enough. The point of the entire exercise seems to be to justify absolute monarchy, and Hobbes spends chapter after chapter detailing all the various rights and prerogatives with which the sovereignty is endowed. In *De Cive*, he opines that "monarchy has more advantages than other forms of commonwealth" (DC Pfc.22), and so it must be by the logic of his argument: other forms of sovereignty are more prone to collapse into factionalism and the advance of private interests over the public good (L 19.4, 131). By the end of his career in *Behemoth*, Hobbes is writing a Royalist account of the English civil war, paying particular attention to the injustice of the constraints placed upon the king by the commonwealth in its refusals to grant requisite funds. One is easily led to the conclusion that "the general system . . . was all about the king: his rights, his power, and the possible limits of his power" (Foucault, 2003, p. 27).

The problem is that the evidence for this interpretation lies entirely on the surface. Beneath the surface, something substantially different is at work. A juridical focus on the king presents the culmination of a lineage that dates at least as far as Aristotle's discussion of the lawgiver, whose mastery of the political art is supposed to assist in bringing about the natural motion of people into the *polis.* It is precisely this lineage that Hobbes contests.[3] Even the passage in the preface to *De Cive* offers a glimpse at this result; after confidently announcing the superiority of monarchy, Hobbes parenthetically takes the ground underneath it away: "I confess that [this is] the one thing in this book not demonstrated, but probably set forth."[4] The argument in the body of the text will even "ignore certain arguments which present *Monarchy* to us as the preferred form, because they work not by reason but by example and testimony;" the second edition appends an elaborate allegorical reading of Prometheus to bolster the point (DC X.3; cf. L 19.4–8). If political philosophy is demonstrable, as Hobbes says repeatedly that it is, then the proposition that monarchy is the best form of government is not even to be admitted in political philosophy. Hobbes's interest lies, as he says, in the derivation of political power in the first place. Thus, the differences between monarchy, aristocracy, and democracy "do not consist in the diversity of juridical power [*potestatis*], but in the diversity of their suitability to the end[s] of the commonwealth, namely peace and defense" (OL III, 142).[5] How, then, can political philosophy essentially be about the king?

Not only that, consideration of the sovereign requires negotiation of Hobbes's understanding of artificial personhood. Hence, even the *Leviathan* chapter designed to prove the superiority of monarchy is based on the claim that "whosoever beareth the person of the people, or is one of that assembly that bears it, beareth also his own natural person" (L 19.4, 131). The sovereignty may need to be vested in a single agent, but there is no need for that agency to be tied to a specific, natural person.[6] Hobbes may thus advocate a king, and he may even be motivated by a defense of Charles II, but the logic of his argument lies elsewhere.

A more obvious candidate for a political philosophy which is all about the king is one that is almost completely lacking in juridical elements: Machiavelli's.[7] Machiavelli's first advice for the king is to dispense with the theological niceties of the discourse about kings and treat them for what they are, discursive but very real elements of the king's maintenance of power. Thus reduced to its material bases, Machiavelli's dissection of power and his emphasis on the visible person of the prince suggest that it is here that we see most clearly the person of the king, deprived of all juridical appurtenances, charged with his most fundamental task, maintaining the stability of the commonwealth against the disorders with which it is threatened. At the same time, Machiavelli's texts thereby present an unwitting *reductio* of themselves. At the theoretical level, the prince is to be armed in his struggle against the vicissitudes of the natural world by careful study of human history. But this study proves aporetic, as it turns out that human history does not reveal the reasons why its successes have been successful. If anything, study reveals the permanent suspicion that situations then were not like those now, and so the advice to the prince devolves into a general admonition to secure his material power. But even the security of material power adheres to no principle; adopting or not adopting almost any of its possible constitutive elements might or might not confer actual security, with the difference being decided by how their use appears. How their use appears in turn depends on the situation in which the prince finds himself, and so the best advice to the prince is to map out as many of the contours of his situation as possible and maximize his relative advantage within that context. Machiavelli's best examples to follow, as a recent paper reminds us, were *sui generis* innovators described according to myth (Breiner, 2008). In any case, all good things must come to an end, and the very traits which brought one to power in a given situation will undermine that power in the next. In that sense, the person of the king is not the necessary condition for the functioning of power, but one of its biggest problems: not only does nature threaten the principality from without but also the natural attributes of the monarch threaten it from within.

Thus facing the specter of the total collapse of Aristotelian models of political thought against the natural disorders they were designed to stabilize, Hobbes takes the bold move of banishing nature from political philosophy. No longer is man the political animal. He is the desiring animal, but these desires are programmatically reduced to the interplay of competing forces and drives; this interplay can then be regulated by the sovereignty, the job of which it is to reorient these desires in a mutually productive way. Thus the "state of nature" is peculiarly named, as its members—the "multitude"—are not considered naturally at all. They are instead first transformed into the atomic elements of the well-ordered "people," which is created by the sufficient application of governmental technology to reorient their desires around a single, coherent signifier. The "people" can now be considered a person, and they can appoint a representative in the form of a sovereign. But the representative does not have to be a natural person either. At the level of theory, Hobbes thus effects a double reversal of Aristotle. On the one hand, because politics is a productive art, and not to be drawn from nature, it is not a science in the Aristotelian sense. The Aristotelian notion of science has in any event collapsed entirely for Hobbes, as even physics itself no longer admits of demonstrative certainty. On the other hand, this politics is able to achieve a demonstrable certainty, against the infinite variability of human affairs, because its governing logic is geometric. In this regard, the Earl of Clarendon was precisely correct when he complained that Hobbes would "erect an engine of Government by the rules of Geometry, more infallible then Experience can ever find out" (1676, p. 117).

Rather than looking at the monarch on the frontispiece of *Leviathan*, then, and concluding that Hobbes typifies the juridical model of power, it would be better to look at the details of the population living in and through it, and ascribe to Hobbes something like a notion of biopower—power that "exerts a positive influence on life, that endeavors to administer, optimize, and multiply it, subjecting it to precise controls and comprehensive regulations" (Foucault, 1978, p. 137).[8] In other words, it is under the aegis of construction that Hobbes's project becomes both intelligible and specific. After speaking of the possibility of creating an artificial animal, Hobbes concludes of his social contract that

> Finally the pact, which holds together the parts of its [the commonwealth's] body, imitates the divine word *fiat* or *faciamus hominem* brought forth by God in the beginning when he created the world. (OL III, 2)

It is this "let us make man" that marks the innovation and the modernity in Hobbes, signifying as it does the abandonment of the Aristotelian dictum that "the political art does not make men but takes them from nature and uses them" (*Politics* I.10, 1258a22).

7.1.3 Our age is Hobbesian

If Whitehead felt safe in characterizing the European philosophical tradition as a series of footnotes to Plato, then perhaps we should risk a much more dangerous view: that "modern" political thought is mostly a series of footnotes to Hobbes. That is, we moderns, whose assumptions are still very much products of the seventeenth-century revision of philosophy, have taken these constructive principles too much for granted.[9] As David Gauthier puts it with regard to Hobbes, the social contract is our ideology: "our thoughts and actions are to be understood as if we supposed that all social relationships were to be rationalized in contractual terms" (1977, p. 136). In so doing, we have tended to focus too much on our own issues: does Hobbes's response to the fool, for example, satisfy a loosely Kantian definition of normativity, and can it thus count as moral? Could actors in the state of nature form a commonwealth? In reading a number of these elements of Hobbes's philosophy against the grain, and with attention to aspects of his context that make those elements pressing, I hope to uncover a glimpse of what was at stake for Hobbes, in his own self-declared project of putting civil philosophy on the same secure footings on which Euclid had put geometry and on which Galileo and his followers had begun to put physics. This recovery is vitally important for the terrain of our own political philosophy, over which the figure of Hobbes hovers like the sovereign depicted on the frontispiece to *Leviathan*: not quite of us, but not that far removed, either. If I am right that the practical maxim which emerges from the constitutive figuration of Hobbesian rationality as constructive is to advance biopower, then we are still very much the products of a Hobbesian world.

7.2 Overturning Hobbes

Deleuze remarks that "the task of modern philosophy has been defined: to overturn Platonism" (1994, p. 59). In the present context, we can add: the task of political philosophy today is to overturn Hobbesianism. What this task requires is that we affirm biopolitics without affirming *Leviathan*. We no longer think absolute monarchy is the best form to organize the *polis*, even in the reduced Hobbesian sense of a probable argument rather than a demonstration. But turns out that getting past the king did not get us past Hobbes; the king, it turns out, had been inessential at least since Machiavelli. What would this mean?

On the one hand, it seems evident that we must affirm the constitutive power of language. However debates about the possibility of prelinguistic thought are settled, it is apparent that Hobbes was right about the political power of language and its capacity to order our affects. Reversion to a premodern understanding of language as part of an effort to overcome Hobbes would be to embrace a series of largely theologically driven metaphysical assumptions; it would be, in other words, to overtly affirm the most problematic aspects of Platonic transcendence against the immanent, constitutive power of language. On the other hand, if the person of the king is inessential to Hobbes, the conceptual structure undergirding that person—the transcendence of ruler over ruled—is not. This structure is most obvious in the way that Hobbes articulates the rights of the sovereign. Accentuating precisely this transcendence, Agamben proposes that its logic be understood as follows:

> It is important to note that in Hobbes the state of nature survives in the person of the sovereign, who is the only one to preserve its natural *ius contra omnes*. Sovereignty thus presents itself as an incorporation of the state of nature in society, or, if one prefers, as a state of indistinction between nature and culture, between violence and law, and this very indistinction constitutes specifically sovereign violence. The state of nature is therefore not truly external to *nomos* but rather contains its virtuality. (1998, p. 35)

The Hobbesian sovereignty, in other words, repeats the conceptual structure of the state of nature/social contract pairing, by casting the bearer of the sovereignty as radically but ambiguously transcendent.

Of greater interest in the present context is another Hobbesian move, one that, as I have argued, Hobbes himself indicates is of utmost importance: the distinction between the multitude and the people. The figure of the multitude stands for axiomatic disorder, for a collection of bodies that cannot be brought into the *polis*. The constitutive social contract, then, constitutes the *polis* as such by transforming the "multitude" into the "people." It seems clear that the theoretical justification for this move is not so much to be found in Hobbes's political philosophy as in his geometry. In particular, it lies in Hobbes's adoption of the law of homogeneity, the expression of the Greek reduction of mathematics to countability. As I have indicated, Hobbes invokes this law explicitly in his attacks on symbolic algebra, and here we can see its general application. Recall that the fundamental problem with symbolic algebra is that it fails to specify the ontology of its referents; "this putting off an unit sometimes for one line, sometimes for one square, must needs mar the reckoning," and "it can be no otherwise when you so apply arithmetic to geometry" (EW VII, 64). By analogy, if it is true that our desires are radically heterogeneous, then the failure to apply homogeneity to the referential field of the *polis* will similarly mar the reckoning about it. Natural difference, which Hobbes's reduction of the intellect to bodily imagination necessarily affirms at the individual or even subindividual level, must nonetheless not be admitted into the conceptual field of political philosophy. Transcendence, at this level, must be preserved, for it is transcendence that enables philosophy, not immanence.

This result is striking, for it indicates that Hobbes's two most prominent theoretical commitments are pulling in opposite directions. His nominalism straightforwardly tells him that power can only be understood on a plane of immanence. But his geometry tells him that such a plane of immanence is incomprehensible without a prior ordering of the elements of that plane into ontological homogeneity; that is, it is only comprehensible by way of a rejection of immanence in favor of transcendence.[10] In other words, the insight of the constitutive power of language is both undermined and stabilized by Hobbes's resistance to the idea that mathematics be fully understood as a language. Some evidence of this resistance, of course, occurs in Hobbes's efforts against metaphor and imprecise speech, but more fundamental and convincing evidence of it is found in the geometry, as for example in his violent rejection of imaginary numbers in algebra. Hobbesian geometry is never about symbols; it is always a matter of univocal referentiality, stabilized with reference to something outside itself. Hobbesian political sovereignty thus constructs a juridical regularity in the state of nature to solve the Aristotelian problem of natural diversity. It is this maneuver that enables the strange status of the sovereign to which Agamben draws our attention. It is also this maneuver, and the tension between nominalism and classic geometry that it embodies, that grounds the fundamental conceptual instability of the *polis* generated by the Hobbesian apparatus. The problem is that the images of his theoretical commitments always risk turning into one another. On the one hand, the productive arts enabled by the *Leviathan* always seem to risk collapsing into a homogeneous, gray despotism. On the other hand, the stability that enables those arts always itself seems to risk collapse into the unregulable anarchy of the state of nature. Most of the theoretical energy of the chapters in *Leviathan* dedicated to what one might call the art of governance is aimed at negotiating this tension, at keeping the *polis* afloat despite its unstable ballasting. Thus the role of institutions like the judiciary and the university become both increasingly relevant and increasingly difficult.

What might all of this mean for a task of overcoming Hobbes? It appears that Hobbes banished nature from the *polis* in name only; thrown out the front door, the classical concept of nature as a nomological ordering returns in abstract form through the window of geometry. One way to pursue our task, then, would be to reject the myth of wholeness and homogeneity enabled by that geometry. We need to reject the notion that our representations of the social order must be based on some sort of ontologically fundamental natural order. That is, we need to pick the nominalism over the geometry; in so doing, we will be allowing ourselves to attend to the embodied human differences that Hobbesian materialism refused to recognize. In other words, to overcome Hobbes, we may need to radicalize him. Aristotle was right: insisting on too much precision in political philosophy is a bad idea, grounded in a refusal to countenance the diversity of human political life.

It will be no accident that the effort to empower the multitude is central to efforts to overcome Hobbesianism, since it is the multitude which first bears all the weight of natural difference and complexity, and which is then consigned by Hobbes to illegibility. Hardt and Negri, for example, align the question of the multitude with the sense of monstrosity. They write that

> We need to find the means to realize this monstrous power of the flesh of the multitude to form a new society. . . . We need to write a kind of anti-*De Corpore* that runs counter to all the modern treatises of the political body and grasps this new relationship between commonality and singularity in the flesh of the multitude. (2004, pp. 193–4)[11]

I want to close simply by noting a difference in emphasis in what might first appear to be a dictum common to the two seventeenth-century moderns who had the most to say about the multitude.[12] The first is from Hobbes, who writes in one of his mathematical treatises, "I know that things which are beautiful are difficult. But the converse—that the difficult at the same time is also beautiful—is not thereby stated." (OL IV, 89). Framed by this warning, the Hobbesian geometric reconfiguration of political thought emerges as a strategy for managing complexity, since it is, on Aristotelian grounds, precisely the complexity of human affairs that limits the precision of political thought. The second comes from Spinoza. In closing his *Ethics* and its discussion of how to move past such strategies, Spinoza suggests that "for how could it happen that, if health [*salus*] were at hand, and could be obtained without great labor, that it would be neglected by almost everyone? For all things excellent are as difficult as they are rare" (E5P 42 SCh).[13] "*Salus*" refers equivalently to the health of people and of commonwealths; our difficulty is to ensure the complexity of our own efforts to think through and against this Hobbesian strategy for dealing with complexity. To reduce Hobbes would be to do with Hobbes, not against him.

Notes

Chapter 1

[1] Qt. in Mintz, 1962, p. 56. Mintz's remains the most comprehensive catalogue of contemporary English responses to Hobbes. Two cautions regarding its use are in order. First, response to Hobbes was not unequivocally negative, particularly on the European continent: see Malcolm, 2002b; Rogers, 1988; and Skinner, 1966. Second, Mintz concludes that the main charge against Hobbes was atheism. As I argue in Chapter 2, Hobbes's opponents also had specific methodological concerns which can be occluded by focusing too much on the most fluid of seventeenth-century pejoratives, "atheist."

[2] Marx also suggests that, in Hobbes, "sensibility loses its flowers and becomes the abstract sensibility of the geometers" (1927, p. 305).

[3] I am here repeating the assessment of Bray, 2007.

[4] Levy 1995, p. 35. Charles William Maynes (1995) singles it out in his critique of "The New Pessimism" in an article of that title.

[5] I will not pursue the point here, but Descartes' role in the constitution of modern political subjectivity has been largely and unjustly ignored. For evidence that it is nonetheless important, see a pair of papers by Timothy J. Reiss: Reiss, 1991 and 2004.

[6] The assaults on a caricatured Descartes are often launched by self-declared "postmoderns." For the thought that the postmoderns are much closer to the moderns than they think, see Lachterman, 1989, chapter 1. My own work is heavily indebted to Lachterman.

[7] Foucault quips that "the philosophy of right subsequently rewarded Hobbes with the senatorial title of 'the father of political philosophy'" (2003, p. 99). The following citations are exemplary of the range of ways in which Hobbes is said to have initiated modern political thought: Gauthier, 1988 ("Hobbes's achievement in being the first systematically to construct such a theory [alienation-based social contract] makes him the true parent of rational morality and politics," p. 126); Sacksteder, 1981b (Hobbes "is with like ambiguity both a modern and an ancient . . . He ushers in most of the movements we call 'modern' when we consider the sweep of western thought. Yet his training and his pretensions are classical," p. 461); and Shulman, 1989 (study of Hobbes lets us ask "what is the relationship between the ongoing rationalization of state power and yet our increasingly theatrical relationship to it . . . Hobbes writes the script people have followed as they

have authorized modernity" (p. 393; a list of several ways in which much of modern statecraft theory is already in Hobbes follows on pp. 401–2).

[8] This point is made particularly clearly in Breiner, 2008.

[9] I am following the discussion in Rahe, 2007, which both cites the passage, and provides the Lucretius reference.

[10] It should go without saying that "Aristotle" and "Aristotelian" are not univocal terms, a point which I will evidence by referring frequently to late scholastic sources. By the seventeenth century, "Aristotle" had become a convenient shorthand for scholasticism in general; Hobbes thus follows contemporaries such as Gassendi and Descartes in this practice. For some of the complexity surrounding the early modern reception of late scholasticism, see, for example, Ariew, 1999; Des Chene, 1996; Leijenhorst, 2001; and Paganini, 2007. For an example of the Hobbesian rhetorical amalgamation of Aristotle and the scholastics, see, for example, L 46.11, 461. Hobbes elsewhere and more sympathetically distinguishes Aristotle from the "thousands who followed him" (OL V, 359; cited in Schuhmann, 2004a, p. 218). It is a matter of some contention how much Aristotle Hobbes actually read, and how much his polemic should be taken as being against late scholasticism. See Schuhmann, 1990 and 2004a. For an attempt to grapple with the complexity of medieval usages of "Aristotelian," see Nederman, 1996, which includes a survey of previous such attempts.

[11] The literature discussing claims such as these is vast. I will discuss geometry in what follows. On developments in science, Dear, 2001, offers a helpful, recent survey. For broader cultural developments, see, for example, Reiss, 1982 (examining utopian fictions) and 1997 (for general discussions of method). I will refer to other studies where appropriate.

[12] In this regard, my general debt to Foucault's genealogical work should be sufficiently evident as not to require marking again in a general way. Although my debt to the work of Quentin Skinner is considerable—indeed, it is perhaps not too great an exaggeration to say that Skinner's careful work has made possible studies such as this one—I am not primarily motivated by an effort to uncover what Hobbes's "intentions" were, and the contextual picture I draw here will transcend considerably the local boundaries that Skinner tends to draw. Full discussion of Skinner's methodology and its origins in speech act theory is well beyond the scope of this introduction. Skinner's (1988a) initial method article is indispensable. For useful caution regarding Skinner's work on the points just mentioned, see, for example, Femia, 1988; Goodhardt, 2000; Keane, 1988; and Schuhmann, 1998. Skinner addresses many of these objections in 1988c. Finally, Tully, 1988, suggests that Skinner should be read as producing a Foucauldian genealogy, with republican humanism as his counter-ideology.

[13] I will substantiate this claim more fully in the following chapters. For exemplary critiques of the (unfortunately still too common) view that Hobbes's political thought somehow is not on the same page as the rest of his thought, see Watkins, 1965; Goyard-Fabre, 1995; and Zarka, 1987: "the entire organization of knowledge is suspended from metaphysics, whose sense will fundamentally engage that of ethics and politics" (1987, p. 14; see pp. 11–26 more generally). A similar reductionist position, evident (perhaps) in Shapin and Shaffer's *Leviathan and the Air Pump* (1985), argues that Hobbes's political commitments generate first his

political philosophy and then his scientific/mathematical writings. The problem with this view is that it is either trivially true or unhelpful. That Hobbes would not consciously write theory that undermined his political beliefs is probably trivially true, but leaves aside questions about whether Hobbes fully understood the full consequences of his theory, as well as begging the question of which of these elements of his thought came first. Suppose the theoretical commitment generated the political beliefs? Furthermore, what is at stake in the demand for a priority between them? The view can be unhelpful when it reductively demands that the writings of a thinker such as Hobbes can have only one context. If the present book proves useful in combating reductionist approaches to Hobbes scholarship, then it will have fulfilled one of its purposes.

Chapter 2

1 For sample surveys of the terrain from various standpoints, see Ashcraft, 1986; Hill, 1997; Reiss, 1992, pp. 10–69; Skinner, 1978; and Tuck, 1993.

2 For discussion of this passage and of *Leviathan*'s relation to Plato more generally, see Schuhmann, 2004a.

3 The full context of the passage is worthy of note: "seeing, *Philautus* [=Hobbes; see below], these same *English* men are most of them a Company of clownish and disingenuous dunces; the *Ecclesiasticks* having no breeding, the *Philosophers* having not talked of motion, and the *Mathematicians* being much o'rerun with the *scab* of ignorance, pride, and *symbols*; the best way will be to draw up your opinion concerning a disjunctive proposition (you may do it in half a sheet) and send it beyond Sea: where . . . *you are much read, understood, and admir'd*: and if you can get so much as one subscription to your Paper, by anybody that knows what belongs to Logick . . . then it shall *rain* or *not rain*; and *Socrates* shall *dispute* or *not dispute*, whenever *Philautus* pleases" (54). One may thus note that the prohibition against studying continental philosophy in England is over 300 years old. From Philautus's response, that "*Logick* is the Mother of all Lyes" (55), we learn that the prohibition's content ("continental philosophy is illogical") has remained constant over that time.

4 Both Clarendon and Hobbes were part of the exiled Royalist community in France. For details of their falling out, which is apparently what instigated Hobbes's return to London in 1652, see Tuck, 1993, pp. 320–5. By the 1676 *Brief View*, Clarendon observes "Mr. *Hobbes* his very officious care that *Cromwell* should not fall from greatness" (60). The charge, based on the fact that *Leviathan* appeared during the interregnum and forbad rebellion, was common among those who did not accuse Hobbes of empowering the king at the expense of his subjects. Quentin Skinner suggests that this charge is premised on a misreading of Hobbes's understanding of "liberty" (1990, p. 145). For some of the difficulties in easily pigeonholing Hobbes into "royalist" or "parliamentary" camps, see Curran, 2002 and Hoekstra, 2004.

5 My source for Philolaus is Carl A. Huffman's introduction to his collection of the surviving fragments in his *Philolaus of Croton: Pythagorean and Presocratic*. That Eachard would call Hobbes "Philolaus" may have been peculiarly apt: Huffman

suggests that "there is no evidence that Philolaus was a serious mathematician. Instead Philolaus wants to apply basic mathematical ideas to philosophy in the same way that Plato does in the *Timaeus* and elsewhere" (1993, p. 10).

6 Fragments 7 and 17 (Huffman, 1993, pp. 226–7, 215) are the sources of this attribution. The details of Philolaus' supposed system are via Aristotle and his pupils (Huffman, 1993, pp. 241–2). To suppose that Philolaus's comment in F7 about a "hearth" at the center of the cosmos had much to do with the "sun" as referred to by Copernicus requires considerable interpretive calisthenics: see Huffman, 1993, pp. 243–61.

7 For discussion of this reference, see Huffman, 1993, pp. 14–15; 28–35, taking it as evidence that Aristotle had a book by Philolaus—"likely the first book written by a Pythagorean" (15)—at his disposal when he composed the passage, and discussing the reason for Aristotle's odd use of the locution "so-called Pythagoreans."

8 Another possible reference is in the *Philebus*, where Socrates refers to a "gift of the Gods," passed on by the ancients in the form of a saying that "all things . . . that are ever said to be consist of a one and a many, and have in their conjunction of limit and unlimitedness"—and that the correct delimitation of intermediates "makes all the difference between a philosophical and a contentious discussion" (15c–17a).

9 At *Lives* 8.85, cited and discussed in Huffman, 1993, pp. 12–13. Schuhmann (2004a) suggests that one of the few Platonic texts with which we can assume Hobbes was familiar was the *Timaeus*. The three references here are not the only ones in ancient texts; for a full catalogue and discussion, see Huffman, 1993.

10 I owe this reference to Linda Shenk. For the manuscript and performance, see Vickers' notes to the text (Bacon, 1996, pp. 535–7). The text should be read for connotations of the term "Philautia"—since it was performed and survives only in manuscript, it would be impossible to attribute knowledge of it to Eachard. The attitudes to philosophy and statecraft are clearly Baconian.

11 There are at least three other English references which, however, seem less important in the present context. First, in Harrington's *Oceana*, Philautus de Garbo, an "heir apparent unto a very noble family," speaks against agrarian laws having to do with inheritance and land distribution (101–4). Second, a phenomenally successful pair of prose romances by John Lyly, *Euphues: The Anatomy of Wit* (1578) and *Euphues and his England* (1580) depicted in the character Philautus a bachelor living in Italy who (by the end of the second novel) had returned to England and married. Finally, in 1615, Barnaby Rich's *Anothomy of Ireland* had the character Philautus argue that Catholic Ireland needed to be subjugated by force, as the natives were so degenerate that "this savage maner of incevylyte" could only be tamed by overwhelming force. For Rich, see Hadfield, 2001 and Pincombe, 2002 (which also provides other sixteenth century references for the name).

12 Hobbes's admiration of Mersenne was unqualified. Skinner concludes that Hobbes's contact with Mersenne was the most decisive element in his growing interest in natural science (1996, p. 252). In his verse autobiography, Hobbes says that "Mersenne was the axis around which every star in the world of science revolved [*Circa Mersennum convertebatur ut axem Unumquodque artis sidus in orbe suo*]" (qt. and trans. in *ibid.*). Eachard's reference is to commendations by Gassendi and Mersenne included in *De Cive*'s third edition, of which Hobbes wrote to Sorbière that "in order to give the printer hope that the book will sell, I need to have it

praised by people who are important and whose word can be believed. Accordingly, both M. Gassendi and the reverend Father Mersenne praised it excessively" (HC 40, 126–7). Despite disagreements, Hobbes also held Gassendi in high regard. See Schuhmann, 2004b.

By the *Second Dialogue*, Eachard has inserted a mock "Bookseller to the Reader," which includes such lines as: "Hast thou a Wife and Children, and are they dear to thee? here's a book for that dear Wife, and for those dear Children. For it does not only sing, dance, play on the Lute, speak *French*, ride the great Horse, &c. but it performs all Family duties. It runs for a Midwife, it rocks the Cradle, combs the Child's head, sweeps the House, milks the Cows, turns the Hogs out of the Corn, whets Knives, lays the Cloth, grinds Corn, beats Hemp, winds up the Jack, brews, bakes, washes, and pays off Servants their Wages exactly at Quarter day; and all this it does at the same time, and yet is never out of breath" (1673, sig. a2, 1r, emphasis in original).

13 The passage is: "since Mr. Hobbes by affected garbs of speech, by a starch'd Mathematical method, by counterfeit appearances of novelty and singularity, by magisterial haughtinesse, confidence and the like had cheated some people into a vast opinion of himself, and into a beliefe of things very dangerous and false; I did presume, with your Graces pardon, to think his writing so fond and extravagant, as not to merit being opposed in good earnest: and thereupon I was very loath to give them too much respect, and add undue weight by a solemn and serious confutation" (1958, p. 3).

14 For recent applications of this thesis to Hobbes, see Bray, 2006 (showing how the constructive capacity of reason generates an anxiety about the future that only the commonwealth can productively channel) and Sacksteder, 1982 (tying construction to personhood and authorship). For the argument that modern philosophy generally should be understood as governed by "construction" or *poiesis*, see Lachterman, 1989 (focusing principally on Descartes and Kant, though he had originally intended also to write a text about Hobbes (p. xiii)). For a survey of "maker's knowledge," see Pérez-Ramos, 1988, pp. 48–62 and 186–9 (discussing Hobbes). A separate discussion might take up the structure of Hobbes's works: *Elements of Law*, in particular, seems clearly to adopt the geometric method as Hobbes would have encountered it in the Mersenne circle (Baumgold, 2004). Why later works—*Leviathan* in particular—seem to drop this structure is a matter of some controversy, which has recently focused on Quentin Skinner's (1996) claim that Hobbes became increasingly convinced of the need for rhetorical persuasion, in addition to scientific or geometric clarity (for skepticism, see Schuhmann, 1998). The resolution of this structural question is distinct from the claim I will pursue, which is that the governing *substantive* assumptions of Hobbesian thought are *poietic* or constructive.

15 *Naturae Logicae* I.II, p. 2B, in Zabarella, 1597.

16 For a good discussion, see Zarka, 1987, pp.193–214, emphasizing both the nominalist connection and principle of intelligibility. For Hobbes with and against late scholasticism, see also Leijenhorst, 2001, pp. 203–7.

[17] This is because the products of the senses and memory, "which are given at once by nature, not acquired by reason," are not themselves philosophy (DCo I.2; OL I, 2).

[18] For a discussion of these two definitions, uniting them by a concern both for right reason and phenomenality, see Lee, 2004, pp. 103–13.

[19] Cf. Sallis, 1996, for the ways nature limits political thought in ancient texts.

[20] As Zarka puts it, "the political foundation of a juridical code of the state is substituted for the lost ontological order" of Aristotelian nature (1987, p. 25). On these points, and on Hobbesian definition as generative, see also Gauthier, 1997; Goyard-Fabre, 1995; and Watkins, 1965, pp. 40–1.

[21] I have attempted, where possible, to use gender neutral language, although I (of course) do not "update" primary texts by making them gender neutral. It is clear that, for us, consideration of the *polis* is not the consideration of men alone, and should include consideration of sexual and gendered differences.

[22] This argument disappears from the Latin text. Both editions suggest an inability to distinguish one action of many people from the many actions of a mere multitude (OL III, 82).

[23] The distinction also occurs in the *Metaphysics*, where Aristotle writes that "art is the principle in a thing other than that which is generated, nature is a principle in the thing itself, and the remaining causes are privations of these" (1070a9). The reading here follows Schummer, 2001, which argues that Aristotle's point is not to try to say that nature in every case imitates art—that interpretation, and the corresponding ontological priority assigned to nature, was a later development. For Aristotle, according to Schummer, the nature/art distinction was perspectival, rather than ontological, and was only actually used by Aristotle in the discussion of composites—when he came to discuss chemicals in *Meteorologica* IV, it largely disappears from view. These details need not be of concern here, since, as I will indicate, the Hobbesian enterprise is concerned with what we can know about the objects in question.

[24] Coimbra, *In Phys.* 2c1xpl, 200b. I choose the Coimbra (or Conimbricenses) commentaries as representative of the state of late scholastic thinking, and make no claim for their influence on Hobbes. The Coimbra commentaries were widely used and read, and Descartes, for example, would have studied them at La Flèche. For the range of scholastic sources Hobbes might have encountered (direct evidence is lacking, except in the case of Suárez), see Leijenhorst, 2001, pp. 8–9. For Hobbes's mature view of motion, see Leijenhorst, 2001, pp. 187–203. Leijenhorst's book is by far the best study of Hobbes's relation to scholastic physics, and I am substantially indebted to it. For further discussion of the scholastic understanding of motion, and of early modern rejections of it, see Des Chene, 1996.

[25] *Physics* 201a10–11. For Scholastic translations and glosses, see Des Chene, 1996, pp. 26n11.

[26] OL I, 176; emphasis removed. Cited in Leijenhorst, 2001, p. 195. Cf. also Leijenhorst, 2007, p. 96, discussing the centrality of inertia (L 2.1, 15) for the rejection of teleology. Spragens, 1973, emphasizes Hobbes's abandonment of teleology,

though that study falls into overgeneralizations about "Aristotelian" and "seventeenth-century" philosophy.

[27] On these points, see Schummer, 2001 and Newman, 1989. For a recent demonstration of this instability, where Aristotle has to invoke mysterious, self-moving automata—understood on the model of machines, and thus *techné*—to explain sexual reproduction, see Bianchi, 2006.

[28] Philostratus says that these statues were "like those in Homer's poem." The reference is to *Iliad* 18.374–9, where Hephestus "was fashioning tripods, twenty in all, to stand around the wall of his well-built hall, and golden wheels had he set beneath the base of each that of themselves they might enter the gathering of the gods at his wish and again return to his house, a wonder to behold." Aristotle, also cited by the Conimbricenses for his reference to the Daedalus statues (*Politics* 1253b35), refers to the Homeric tripods in the same passage.

[29] For this, and speculation on its possible influence on Descartes, see MacDonald, 2002, pp. 438–48.

[30] For these, see Newman, 1989.

[31] Cited in Pettit, 2008, p. 57. Tralau, 2007 makes the case that the Leviathan itself should be viewed as such a monster. Cf. Donna Haraway's invocation of the effects of cyborgs, as those beings which illicitly violate the artificial/natural boundary: "The certainty of what counts as nature—a source of insight and promise of innocence—is undermined, probably fatally. The transcendent authorization of interpretation is lost, and with it the ontology grounding 'Western' epistemology" (1991, pp. 152–3).

[32] For a thorough treatment of the complexities of the nature/art distinction as it maps onto the *Politics*, see Nederman, 1994.

[33] The thought that the state is artificial was a scholastic commonplace (on this see, e.g., Zarka, 1987, pp. 226–8, suggesting that Hobbes's modernity is grounded in his ontology). The specificity of Hobbes's development of that thought was uncommon, as will become apparent. For *Leviathan* as monstrous, see Stillman, 1995. For the Biblical *Leviathan*, see Mintz, 1989. I pursue the thought that the Job reference functions allegorically in Hull, 2002, p. 20. Bredekamp, 2007, identifies a precise possible source for the Leviathan as monstrous, as depicted in the frontispiece: the hermetic *Alscepius* text, which refers to "man the creator of gods," *viz.* "statues that have life breathed into them, full of spirit and *pneuma*, that accomplish great and mighty deeds, statues that can read the future and predict it through priests, dreams and many other things, which weaken and heal men, create sadness and joy for every individual according to his merits" (qt. p. 34).

[34] For the textual history of these maxims and of Hobbes's formulation, see Tricaud, 1969.

[35] I am referring to Theodore Goulston's Latin translation: "Quae enim praeter natura sunt (ut monstra, &c) non oportet éxactè disputare . . ." (1619, p. 55). Goulston's facing-page edition was the first widely available in England; Hobbes evidently knew it.

[36] Aristotle refers to this passage again at *Politics* III.6: "men are by nature political animals, and for this reason, even when they have no need of each other's help,

they desire no less to live with each other; and, moreover, common expediency brings them together to the extent that it contributes to the good life of each" (1278b19–23). For more discussion of Hobbes's rejection of the "political animal," see Schuhmann, 2004a, p. 210.

[37] That is, the distinction between violent and natural motion is otiose for Hobbes. Citing texts as early as *De Motu*, Leijenhorst points out that, for Hobbes, "when motion is no longer considered teleological, distinguishing between violent and natural motion becomes meaningless" (2001, p. 194). When Hobbes's arch-enemy John Wallis (see ch. 3) called "mechanics" the "geometry of motion," Hobbes had no objections. On these points, see Jesseph, 2006. I will not further trace here the reconfiguration of mechanics from a motion *praeter naturam*, through Galileo, and to the seventeenth-century moderns. Among the background issues which would emerge in such a lineage would be the question of the certainty of the mathematical sciences, the relative status of mixed and pure sciences, the elevation of mathematics at the hand of Clavius in particular, and the importance to all of this of developments in military technology. For a survey of the pseudo-Aristotelian text, see Rose and Drake, 1971.

[38] For some of the difficulties in "equality" as a postulate of early modern political philosophy, see Waldron, 2002. Waldron's reading of Locke—roughly, that for Locke, we are all equal before God insofar as reason leads us to recognize God's existence and rule over us—allows one to underscore the extremely minimal role that such a theological apparatus plays in Hobbes.

[39] Grotius should perhaps be seen as somewhat liminal. Although his discussion is wide-ranging and draws on numerous sources, he nonetheless asserts the priority of reason over doxa: he will "undertake to treat the parts of the natural and unchangeable philosophy of law, after having removed all that has its origin in the free will of man" (*De iure belli* . . . Prolegomena, §31). For this sense of critique as purification in the modern period (and its subsequent reversal in the nineteenth century), see Röttgers 1975.

[40] The Latin text seems to ontologize the requirement: the ninth law of nature is that "homines omnes inter se natura aequales esse" (OL III, 118). That said, the epistemic sense remains in the prior clause: "atmittenda ergo est aequalitas . . ."

[41] See also *Politics* I.1: it is a *sine qua non* of the political that "there must be a union of that which by nature can rule and that which [by nature should be] ruled, for the sake of their preservation; for that which can foresee by thought is by nature a ruler or by nature a master, whereas that which [cannot foresee by thought but] can carry out the orders with the body is by nature a subject or slave" (1252a31–4), and I.5: "Those differing from others as much as the body does from the soul or brutes do from men (they are so disposed that their best function is the use of their bodies) are by their nature slaves, and it is better for them to be ruled despotically" (1254b15–20). For discussion of Hobbes's rejection of this passage, see Schuhmann, 2004a, pp. 210–14.

42 The reasons adduced here do not take into account the rhetorical tradition, with which Hobbes was obviously familiar. Hobbes's invocation of techniques for persuasion is treated in great detail in Skinner, 1996, and there is no need to repeat Skinner's analysis here. My point is simply that there are also reasons within scholasticism to treat these issues, and, as I will indicate, those reasons point to one of Hobbes's abiding concerns: the status of universals.

43 I also do not wish to engage in the Straussian question of whether Hobbes's texts are esoteric, and are deliberately designed to lead to unbelief. Strauss is heavily influenced by his reading of Maimonides, for whom esotericism is a stated desideratum ("my purpose is that the truths be glimpsed and then again be concealed, so as not to oppose that divine purpose which one cannot possibly oppose and which has concealed from the vulgar among the people those truths especially requisite for His apprehension" (1963, I, 6–7)). The question of esotericism in Hobbes is usually broached with respect to atheism. Here I will bracket that question (but see ch. 4, n24 and accompanying text), as I do not think there is any reason to suspect esotericism about Hobbes's methodological pronouncements.

44 Analysis and synthesis also do not mean exactly the same thing to a seventeenth-century thinker as do induction and *syllogismos* to Aristotle. For a representative discussion of method, see, for example, Descartes' clarifications in the *Second Replies* (AT VII, 156–7; PWD II, 110–12). I will confine my discussion here to Hobbes.

45 For some discussion of Machiavelli's tremendous cultural impact, see Reiss, 1992, pp. 42–69, noting the European sense of a "fusion of linguistic failure and actual catastrophe" (44) by 1600. For Machiavelli in England, see Raab, 1964. The reading of Machiavelli presented here is intended as a thumbnail sketch, and by necessity leaves many interpretive questions open. In lieu of a detailed discussion, let me indicate a pair of my substantial interpretive debts: (1) Whether or not one agrees with his final conclusions, Vatter (2000) strikes me as a tremendously productive reading of Machiavelli. As will become apparent below, I think that Aristotle is less committed to "form" than Vatter takes him to be; that said, Vatter's emphasis on events, coupled with his clear embeddedness in Reiner Schürmann's work on *arché* in Heidegger, allows one to underscore the extent to which the possibility and contours of practical reason are at stake in Machiavelli. (2) The difficulties of *The Prince, qua* text, need to be emphasized. On this, Greene (1986) is an essential starting point. I will reference other discussions below, as appropriate.

46 Gianfranco Borrelli, emphasizing Hobbes's break with theorists influenced by Bodin, suggests that, for Hobbes, "the dispositives of prudence act first as an aid to techniques of force and fraud. Far from facilitating the action of the sovereign, these techniques on the contrary constitute one of the principal causes of the dissolution of the state" (1997, p. 21). For a general discussion of prudence in Hobbes, emphasizing in particular the various reasons why it is prone to failure, see Vanden Houten, 2002. Hanson, 1993, emphasizes the extent to which it is not prudence *per se* that is the problem, but prudence unhinged by erroneous or opaque principles.

[47] Hobbes has something similar in mind when he remarks in *Leviathan* that "nor does it alter the case of Honour, whether or action (so it be great and difficult, and consequently a signe of much power,) be just or unjust for Honour consisteth onely in the opinion of Power" (L 10.48, 66); the heathens, for example, intended to honor their gods when they depicted them in raping and pillaging (*ibid.*), and absent the laws of a commonwealth, both piracy and duels are honorable (L 10.49, 67).

[48] This is a departure from standard Italian advice books for princes; see Skinner, 1998b, pp. 424–5 and 432ff. The emphasis on appearances is repeated again when Machiavelli analyzes the current situation in Italy: "the actions of a new ruler are much more closely observed than those of a hereditary ruler" (P XXIV, 83). See also *Art of War*: "what above everything else keeps the army united is the reputation of the general; this comes only from his ability, because neither blood nor rank ever gives it without ability" (W 698; cf. also the discussion of "fear of arms," W 606). Machiavelli analogizes the situation of a new prince to a new general in *Art of War*, W 722.

[49] I discuss this passage and the argument behind it in more detail in Chapter 5.3.

[50] There is some controversy as to whether or not Machiavelli "revives" Aristotle. That debate seems misplaced to me; in its stead, I would like to suggest that Machiavelli's texts show evidence of a critical appropriation of a certain reading of Aristotle. For the affirmation of Aristotelianism, see Pocock, 1975; for a critique, see Sullivan, 1992. For a discussion of the perils of using the term "Aristotelian" without further specification, see Nederman, 1996. Finally, one should note the Thomism of the Savonarolans that Machiavelli opposed; on this, see Colish, 1999, pp. 609–10.

[51] Cf. also: "[virtue] grows by those actions by which it is in the process of coming into being but is destroyed if those actions are not done in this manner" (NE 1105a14–15). In this way, a judgment of virtue must always transcend a given act. Aristotle emphasizes this earlier: "and we should add 'in a complete life,' for one swallow does not make a spring, nor does one day; and so too one day or a short time does not make a man blessed or happy" (NE 1098a18–21).

[52] See *Discourses* III.6: "As to the dangers undergone in executing a conspiracy, they come either from varying the arrangements or from lack of courage in their executor or from a blunder he makes through imprudence or through not completing the thing." Note especially the importance of imagination: "If men have fixed their imaginations for many days on one method and one arrangement, and that suddenly changes, by no possibility can they avoid being completely upset and everything ruined" (*ibid.*).

The rethinking of *virtù* by aligning it with action even produces a reversal at the question of what age group is suited to politics. For Aristotle, in a sentiment echoed by Guicciardini in Machiavelli's time, "a young man is not a proper student of politics, for he is inexperienced in actions concerned with human life, and discussions proceed from [premises concerning those actions] and deal with [those actions]" (NE 1095a4–5). Machiavelli—and recall the parallel between political and military leadership—instead refers approvingly to youth and complains that "men of another age, with their hair white and the blood in their bodies turned to ice, are commonly some of them enemies of war, some beyond correction, believing that the times and not bad customs force men to live thus" (W 573).

[53] Compare Hobbes: on the one hand, Hobbes both rejects the Aristotelian mean as the effect of Aristotle not having a coherent theoretical account of the virtues (L 15.40, 111) and distinguishes between just people and actions (L 15.10, 103–4). On the other hand, Hobbes takes these together to be sufficient grounds to reject contextual analysis altogether.

[54] Thus, "fortune . . . shows its powers where no force has been organized to resist it," but on the other hand, "we are successful when our ways are suited to the times and circumstances, and unsuccessful when they are not" (P XXV, 85). Thus also the *Discourses*' praise of Rome: "if no republic ever produced such results as Rome, there has never been another republic so organized that she could gain as Rome did. The efficiency of her armies caused her to conquer her empire, and the order of her proceedings and her method, which was her very own and discovered by her first lawgiver, caused her to keep it when conquered" (II.1). He follows with a lengthy and detailed discussion of Rome's particular historical circumstances, and its adaptations to those circumstances. For further discussion, see Nederman, 1999, pp. 623–4.

[55] Fabrizio says similarly in *Art of War*, "you have to vary the form of the army according to the nature of the site and the nature and number of the enemy," and he expresses this as a qualification of the "general rule" by which one should normally form an army (W 642). For a different reading, where the relation between acting and adaptation which I am attributing to Machiavelli is attributed to Guicciardini, and Machiavelli's *virtù* "sought to dominate fortune," see Pocock, 1975, pp. 238ff. In the following, I wish to emphasize the order of decisions: first one decides what kind of action is appropriate, then one does that decisively. This "separation of powers" is paralleled in Florentine and Venetian governmental theory, where it became increasingly important that one group choose a series of possible actions, and another select one from those and carry it out. See Pocock's discussion of Giannotti and Contarini (1975, pp. 272–330).

[56] In the dedicatory letter to the text, Machiavelli claims that "I have not found among my belongings anything that I hold more dear or valuable than my knowledge of the conduct of great men, learned through long experience of modern affairs and continual study of ancient history" (P Ep. Ded., 3). I thus think Hariman, 1989, tends to overstate the extent to which Machiavelli breaks with textual commentaries. It seems to me that Machiavelli's importance lies in the way he foregrounds interpretation, and in the way he emphasizes ancient histories as the communicators of events. Both of these aspects would fit Hariman's general thesis. Aristotle, it should be emphasized, had intimated as much: in his discussion of the intellectual virtues in *Nicomachean Ethics* VI, he suggests that "both politics and prudence are the same disposition, but in essence they are not the same" because they apply to different spheres of human activity (1141b24–5). Thus, both resist reduction to a craft or other form of *techné*.

[57] The best summary of these developments is Jardine, 1988, esp. at pp. 686–93. I discuss *regressus*, and Hobbes's response to it, in Chapter 3.2.

[58] In the *Discourses*, Machiavelli advises that a ruler should appoint indigenous governors of newly conquered lands, so that the people will think they are governing themselves, since "cities . . . rest more tranquil and content under a government they do not see" than one "which reminds them daily of their servitude" (II.21, 342). In other words, whatever trouble ensues as a result of new laws the people will blame on the indigenous governor, and not on the conquering power. Here, the ruler's power requires the opposite of what one might expect: not his spectacular visibility, but his complete invisibility. Cf. Foucault: "if it is true that Machiavelli was among the few . . . who conceived the power of the Prince in terms of force relationships, perhaps we need to go one step further, do without the persona of the Prince, and decipher power mechanism on the basis of a strategy that is immanent in force relationships" (1978, p. 97); on the standard sixteenth-century reading, the link between the prince and his principality "remains a purely synthetic one and there is no fundamental, essential, natural and juridical connection between the prince and his principality. As a corollary of this, given that this link is external, it will be fragile and continually under threat" (1991, p. 90).

[59] There has been relatively little scholarly attention to the relation between Hobbes and Machiavelli. For exceptions, see Foisneau, 1997; Lazzeri, 1990; Skinner, 1996, pp. 170–2; and Strauss, 1952. J. G. A. Pocock suggests a parallel between a Machiavellian prince attempting innovation against fortune and actors in a Hobbesian state of nature (1975, p. 165).

[60] See Foisneau, 1997. Foisneau reads Hobbes's discussion of virtue as a response to Machiavelli (rather than a reception of Descartes, as Strauss would have it), and an attempt "to show that the sense which is given to the word 'virtue' depends entirely on the force of reason, and not on the course of [historical] events" (379–80). Thus, he argues that for Hobbes, "in insisting on the determinant role of fortune in the production of action, Machiavelli resolutely ignores rational predictions, and privileges an uncertain logic of accident" (380). The passage is embedded in Hobbes's discussion of the Biblical fool (see ch. 5.2); for this context and Hobbes's general point that it is bad to advocate disobedience, see especially Hoekstra, 1997. Given Machiavelli's discussion of imitation, it is worth noting Hoekstra's emphasis on Hobbes's thought that disobedience inspires others to disobey; for a discussion of the importance of mimesis to Hobbes, see Kahn, 2001.

[61] Richard Schlatter locates an emergent sixteenth and seventeenth-century cultural preference for Thucydides (in writers such as Bacon) at the expense of Livy in particular (1945, pp. 355–6). For Hobbes's early writings as fitting a Renaissance view that he later abandoned, see, for example, Watkins, 1965, pp. 17–22.

[62] On this, see Bredekamp, 2007 and Malcolm, 2002g. Bredekamp points out that Hobbes seems to precisely refer to his own philosophy as such a perspective glass (p. 42, quoting L 18.20, 129). For a discussion of the image as inherently ambiguous, and of how that ambiguity symbolizes the Leviathan's strange transcendence relative to the *polis*, see Tralau, 2007.

[63] Examples could be multiplied; Charles D. Tarlton counts over forty cases where Machiavelli commends murder as a political strategy. As he notes, "opponents, whose counter-strategies might otherwise have to be anticipated and countered, drop from sight, dead and gone. Populations, whose loyalty might otherwise have to be engineered politically, are slaughtered and their towns plowed under." Hence, "the political science, which starts so confidently and gives off such an aura of definiteness and clarity, comes ever more frequently to collapse into indeterminacy and into the attitude that 'it all depends.' The only solution to the pressing political situations that remain is to crash through them violently, clearing the board and scattering the pieces" (2002b, p. 55). See also D'Amico, 1987, suggesting that for Machiavelli, "the act of violence stages for us that physical, psychological, and social dismemberment that always threatens to reduce human existence to the bestial. When effective as a form of sacrificial ritual, the extraordinary act prepares the way for the creation of an order that will provide justice and security under law" (25).

[64] Dacres is horrified by the entire chapter, and notes that prior to it, had found "not any thing much blame-worthy." Borgia's rise is followed by an inglorious fall, as he fails to support the right candidate in papal succession (on this, see D'Amico, "Machiavelli's Borgia"). Dacres concludes with evident relief that "methinks this example might have given occasion to our Author to confess, that surely there is a God that ruleth the earth" (551). Dacres is usefully discussed in Hariman, 1989, pp. 20–1.

Chapter 3

[1] This is essentially the narrative presented in Jesseph, unpubl. ms., which has the merits of being both informative and entertaining (at one point Jesseph remarks that Fermat "had the good sense to die" before being drawn into refuting Hobbes). Hobbes's declaration of (in)sanity is at OL 4, 387; quoted in Jesseph, 22. Jesseph, 1999b, remains the best (and only major) work on Hobbes's mathematics; a similar account appears in Jesseph 1999a. For the development of mathematics in England more generally, see Pycior, 1997, discussing Hobbes at pp. 135–66.

[2] Eachard elsewhere quips that "by his Logick I profited wonderfully: for it was there (and I must ever acknowledge it) that I first was instructed, to call Logick Computation: and there I learnt how to add and subtract Logically: also how to make use of Triangles, Circles, Parabola's, and other Mathematical instances; instead of *homo, lapis or canis*: and that's, upon my word, what I found there" (1673, sig. a4).

[3] Qt. in Bird, 1996, p. 229. In other words, academic concerns were overdetermined by other issues. Jesseph reads the dispute as primarily grounded in issues of university politics (1999b, pp. 48–72). That discussion should be set alongside the one in Probst, 1993, which emphasizes accusations of Cromwellianism.

[4] HC, 98. The other cause was a dispute over whether Hobbes had plagiarized from Descartes' *Optics*. See HC, 100.

[5] Jesseph, 1999b, p. 350, catalogues the astonishing diversity of Hobbes's mathematical adversaries.

[6] For a quick survey of initial problems with that thesis, see Jesseph, 1999b, pp. 351–2.

[7] On this priority, see, for example, Pycior, 1997, p. 145 and Jesseph, 1999b, pp. 224–46. For a general discussion of synthesis and analysis in the sixteenth and seventeenth centuries, see Dear, 1995. As will be evident, that Hobbesian geometry is subordinate to logic is of central importance in understanding it, a point usefully emphasized in Sacksteder, 1980. See also Malet, 2001, critiquing the thesis that geometry provided a "paradigm" for Hobbes's natural philosophy. Malet suggests that in Hobbes's optics, for example, "when it fails to support causal explanation, Hobbes has no place for geometry, not even as a language of description" (316). The idea that Hobbes is responding to academic skepticism is a commonplace; see the developments of it in Hanson, 1993 (drawing on Tuck); Missner, 1983; and Tuck, 1993. Skinner complains that "the argument has been elevated into an orthodoxy," and that Hobbes "was not primarily responding to a set of epistemological arguments. Rather he was reacting against the entire rhetorical culture of Renaissance humanism within which the vogue for skepticism had developed" (1996, p. 299n32 and p. 9; he is speaking primarily of Hobbes's early writings). Leijenhorst 2001, pp. 11–15, contests the skeptical reading on doxographic grounds. See also Zarka, 1984, for a reading of Hobbes on these points that emphasizes his nominalist empiricism. At the risk of being accused of failing to take a position, I see no reason why Hobbes cannot be responding to several things at once. What I would urge here is that the evidence of his mathematical writings strongly suggests an engagement with questions of analysis and synthesis.

[8] In other words, "it is with the notions of space, body and accident that Hobbes will come to describe phenomenality. . . . However, these notions present no deeper ontological insight than the force of phenomenality itself. Rather, all they afford is a way to trace the forces that give rise to phenomenality so as to make them useful and intelligible" (Lee, 2004, p. 113).

[9] Descartes' *Regulae* would appear to be a target here, as Hobbes language speaks to the impossibility of setting down a certain rule ("*certa statui regula*"). For more on Hobbes's use of analysis and synthesis in this context, see, for example, Zarka, 1987, pp. 74–6 and 168–76, which connects the issue to the collapse of any sense of natural signs. I discuss Hobbes's rejection of natural signs in the following chapter.

[10] See Wallace, 1984, p. 117. For a summary of the scholastic debate, see Jardine, 1988, esp. at pp. 686–93. Watkins, 1965, though generally read as supporting the thesis that Hobbes's methodology is lifted from the Paduans, does point out that Zabarella has reservations about analysis. Dear, 1995, pp. 26–8, points out that successful *regressus* was incompatible with a thoroughgoing nominalism. For early modern implications, see Reiss, 2000.

[11] As Dear emphasizes, none of this is Hume's problem of induction, as "experience" in the Renaissance does not connote a singular, observed event, as it does for Hume. Dear's example is "the effect, or phenomenon, of cold weather in winter. A constant concomitant of that situation is the evening visibility in the northern hemisphere of the constellation Orion. Nonetheless, it might seem implausible to reverse the analysis that identified this constant concomitant so as

to say that coldness in winter occurs *because* of the visibility of Orion. However, it might seem a good deal less implausible if a different constant concomitant were taken as the cause of winter coldness, namely, the fact that the sun in winter is much lower in the sky. The difficulty lies in codifying the procedure whereby causal status is assigned or denied in any given case" (1995, pp. 27–8).

[12] Thus, the commonplace according to which Hobbes straightforwardly adopted "resolutive-compositive method" is false. For a definitive rejection of that interpretation, see Prins, 1990. See also Jesseph, 1999b, pp. 239–40 and 239n37; for the assertion of a greater Galilean influence, see Watkins, 1965, pp. 40–5.

[13] The passage is cited in Jesseph, 1999b, p. 205; and in Prins, 1990, 40n61. Prins also notes EW VII, 308: "egregious logicians and geometricians, that think an induction, without a numeration of all the particulars sufficient, to infer a conclusion universal, and fit to be received for a geometrical demonstration."

[14] The explicit use of "analytica" and "synthetica" does not occur in the English edition (EW I, 316–17).

[15] For evidence that Hobbes studied geometry much earlier than is generally supposed, and that his Euclid was heavily mediated by Proclus, see Schuhmann, 1985. The mediation of Proclus matters because for Proclus, as Perez-Ramos puts it, "the archetypal maker/knower is neither God nor the craftsman or artisan, but the geometrician. The exactness and certainty of mathematical constructs result from their being the soul's own creation which, uncontaminated by sense, cannot deceive himself. The mathematician knows his truths because he has made them." Perez-Ramos adds that this "pattern of thought . . . was to be repeated almost verbatim by Cusa, Mersenne, Hobbes, Vico, and Kant in modern times" (1988, p. 56). Lachterman proposes that "detailed study might show that in crucial respects Proclus is the first 'modern' precisely because of his exaltation of *phantasia* as form-giving" (1989, p. 90).

[16] See chapter 2.2, "Thinking as Construction," for further discussion of this point. For a general discussion of the differences between ancient and modern understandings of mathematics, see Klein, 1968. On the "reversal" of ancient geometry, such that in modern analytic systems, points only have meaning relative to a system of reference, see also Vinciguerra, 1999.

[17] Kepler to Joachim Tanckius, May 12, 1608; original text: *Opera omnia* I, 378. Cited from Pesic, 2000, p. 55. Pesic links the critique of algebra to both Platonic sensibilities in Kepler and his harmonics. The letter is also cited by Reiss, who comments that for Kepler, "all phenomena are thus subject to an unlimited series of interpretations" (1982, p. 142; for Kepler's *Somnium* as liminal, see pp. 140–67). Reiss also cites Cassirer, who claimed that the letter represents "the key to his [Kepler's] entire life's work" (qt. 142, n5).

[18] It should be noted that in both Wallis and Hobbes mathematics will be considered as the paradigmatic *scientia*, a considerable departure from Aristotle. On this, see Mancosu, 1992.

[19] On this immediate referentiality of number in Greek thought, see Klein, 1968. For further discussion of quantity as relational in Hobbes (and thus not the same as numeration), see Sacksteder, 1981a.

[20] For a thorough discussion of the development of the Scholastic understanding of modes, and its development out of an effort to deal with Ockham's critique of Scotus, see Menn, 1997.

[21] "Equinitas sit tantum equinitas—nec est de se una nec plures, nec universalis nec particulares" (Qt. in *Ord.* II, d 3, p 1, q 1, 31). The 1508 Venice edition of Avicenna gives the passage as follows: equinitas "ex se nec est multa nec unum nec est existens in his sensibilius nec in anima. Nec est aliquid hoc potentia vel effectu: ita ut hoc contineatur intra essentiam equinitatis" (1508/1961, fol. 86va).

[22] *Summa Metaphysica* pars 3 disp. 2 §1 q. 3. For a general discussion of the question of rational intelligibility as it intersects with scholastic theological requirements, see Lee, 2002.

[23] See Des Chene, 1996, pp. 367–75, for a survey of the problem and solutions to it. Des Chene suggests that the nominalist approach was not generally followed.

[24] See SL I, 63 and the following paragraphs. For personal supposition in the critique of Scotus, see *Op. Th.* II, 200.

[25] Cf. *Leviathan,* in which there is a discussion of separated essences "to this purpose, that men may no longer suffer themselves to be abused, by them, that by this doctrine of *Separated Essences,* built on the Vain Philosophy of Aristotle, would fright them from Obeying the Laws of their Countrey, with empty names" (L 46.18, 465).

[26] For more on Hobbes's critique of "ontological discourse," see Zarka, 1987, pp. 136–50. Jesseph suggests that for Wallis, Hobbes is guilty of confusing mathematics and physics. It seems reasonable to suggest at this point that for Hobbes, the risk is worth taking in order to avoid confusing metaphysics and mathematics (or, more precisely, to avoid an unnecessarily abstract metaphysics). See 1999b, p. 130.

[27] This point is emphasized both in Pycior, 1997; and in Jesseph, 1999b.

[28] In the Latin edition, the nominalism is even more explicit. For example, the central "he that can tell ten, if he recite them out of order, will lose himselfe, and not know what he has done," becomes "is qui decem verba habet numeralia, nise ordine ea recitet, numerare usque ad decem non potest" (OL III, 26)—someone who has ten *number words,* unless he follows correct order, is unable to numerate all the way to ten.

[29] The claim about political science is in the dedicatory epistle to *De Corpore;* see OL I, nn. Hobbes cites the ancients as his model at EW VII, 188.

Chapter 4

[1] AT VII, 178, my translation; cf. PWD II, 125–6.

[2] Watkins, 1965, 87ff, provides a useful starting place for a possible Hobbesian influence on Leibniz.

[3] I have found the following studies particularly useful: Hanson, 1991, 1993; Leijenhorst, 2001, 2007; Watkins, 1965, pp. 138–62; and (especially) Zarka, 1984, 1987, and 1995. Spragens (1973, pp. 86, 113, 132, 140) frequently alludes to Hobbes's "radical nominalism" without explaining very much what that is, except as a denial of "Aristotelian realism." Zarka summarizes the general interpretive point: for Hobbes, "man is not only a being of desire, but a being of speech," such that Hobbes's political philosophy is not a physics but a "semiology" (1995, p. 51).

Indeed, recent study of Hobbes and language has primarily focused on his conflicted attitude toward rhetoric, the art of persuasion. This has been the subject of several fine, recent studies, of which Skinner, 1996 is the best. For a recent paper emphasizing the importance of the control of language to Hobbes, see Ball, 1995. One might also consult Johnston, 1986 and Whelan, 1981.

4 And not about referential meaning in the twentieth-century analytic sense. From the point of view of a twentieth-century analyst, Hobbes may appear as "the crudest kind of nominalist . . . blind to the 'performative' functions of language," as Ball, 1995, p. 103, puts the objection, prior to answering it. Hacking, 1975, shows the incompatibility between Hobbes and standard, twentieth century theories of meaning. In earlier literature, Hobbes was often accused of "Humpty-Dumpty" nominalism, the view that word meaning reduces to individual will. One corrective to such theories of meaning is speech act theory; it is perhaps for this reason that more recent analytic studies of Hobbes's views on language often cast him as a speech act theorist or as a pragmatist. See, for example, Biletzki, 1997; Parry, 1967; and Bell, 1969. Hungerland and Vick, 1973, similarly emphasize the communicative and public aspects of Hobbes's account of language. Insofar as speech act theory emphasizes the irreducibility of context and actual usage to an account of language, this is both important and correct. However, different views both of signification and of epistemic psychology are compatible with speech act theory: Biletzki, for example, is able to spend very little space on the psychological aspects of signification. It is these elements I will emphasize here. Finally, Sacksteder, 1981b, proposes that most of what passes for the "linguistic turn" in twentieth-century (analytic) philosophy can be detected already in Hobbes.

5 Thus Zarka is able to refer to "the great trauma—not too strong a word—which the Ockhamist theology of all-powerfulness . . . inflicted not only on ontology but gnoseology, but also on morality (by the argument that the law's only foundation is the divine will, which is subject to no obligation) and law" (1999a, p. 16).

6 Importantly, he traces his own view to Aristotle: "the Explication . . . of which, and the like Terms, is commonly in the Schools called *Metaphysiques*; as being a part of the Philosophy of Aristotle, which hath that for title; but it is in another sense; for there it signifieth as much, as *Books written, or placed after his naturall Philosophy*: but the Schools take them for *Books of supernaturall Philosophy*" (L 46.14, 463; see the similar reference at DM I.1). Even more strikingly: at the end of the critique of Aristotelian separated essences, Hobbes adds parenthetically: "it may be he knew [separated essences] to be false philosophy, but writ it as a thing consonant to, and corroborative of, their religion—and fearing the fate of Socrates" (L 46.18, 465). On the genealogy of reading Aristotle as an esoteric atheist, and for discussion of the *De Motu* commentary on Porphyry, see Paganini, 2007. Even today, the materialist Aristotle is heterodox; for some recent work that suggests that Hobbes is on to something, see Barrachi, 2003.

7 For a reading of Ockham emphasizing this ontology, see Alféri, 1989. Because of my emphases here, I will often depart considerably from Loux's (1974b) translations of Ockham's *Summa Lociga*.

8 "Nihil esse universale praeter *nomina*. Natura igitur si res sit universalis, non est" (DM VI.6).

[9] Cited in Prins, 1990, p. 31 n29. Distinguishing Hobbes from Zabarella, Prins notes that "whereas Zabarella considers logic an instrument for the arrangement of notions that, thanks to its metaphysical foundation, enables the scientist to reveal objective reality, to Hobbes logic is a technical science or scientific technology of language by means of which the scientist can construct and consolidate truth itself" (31).

[10] In what follows, I depart from Hacking, 1975; Hanson, 1991 and 1993; Largeault, 1971, p. 192; and Spragens, 1973, p. 144; which all ascribe some form of intellectual intuition to Hobbes. On these points, I am in substantial agreement with Leijenhorst, 2001, pp. 89–97; Leijenhorst 2007; Malet, 2001, pp. 306–7; and Zarka, 1987, pp. 83–182 and 1995, pp. 92–5. Leijenhorst puts the point succinctly: "in Hobbes's cognitive psychology, language thus takes over the role of the Scholastic's active intellect" (2007, 97).

[11] This reading is confirmed with reference to *Leviathan*: "The Imagination that is raysed in man (or any other creature indued with the faculty of imagining) by words, or other voluntary signes, is what we generally call *Understanding*" (L 2.10, 19). In the parallel passage, the Latin text uses "*intellectus,*" the intellectual faculty (OL III, 14). Also, "besides Sense, and Thoughts, and the Trayne of thoughts, the mind of man has no other motion; though by the help of Speech, and Method, the same Facultyes may be improved to such a height, as to distinguish men from all other living Creatures" (L 3.11, 23); the Latin edition repeats the argument and refers to "sense, imagination, and series of cogitations" (OL III, 20).

[12] The comparison to physics is at 1271A. The preceding is a vulgarization of an exceedingly complex debate. For brief summaries, see Boler, 1982; Leijenhorst, 2001, pp. 59–60; and Leijenhorst, 2007.

[13] For further discussion, see Alféri, 1989, pp. 74–87. As he puts it, "this intellection is not the result of a process or an operation. It does not even imply the production of a mental sign. It is concomitant with sensible intuition and exactly overlays the object itself. When I perceive a rose, I think it" (87).

[14] Ockham is thus commonly taken as developing an account of what we would now call an "ideal language." The *locus classicus* of this reading is Trentman, 1970.

[15] See also EL 5.1: "a MARK therefore is a sensible object which a man erecteth voluntarily to himself, to the end to remember thereby somewhat past, when the same is objected to his sense again."

[16] I take this passage as evidence that Watkins is mistaken when he claims that Hobbes's nominalism is inconsistent in that "he sometimes allowed that a common name may stand for something which is *not* individual and singular—for a characteristic property or (as he called it) an *accident* which may be shared by many individual things" (1965, p. 144; emphases in original). The most that one might say is that *similar* motion is shared—but that is a far cry from any sort of real accidents. Cf. EL 2.10: "whatsoever accidents or qualities our senses make us think there be in the world, they are not there, but are seemings and apparitions only. The things that really are in the world without us, are those motions by which these seemings are caused. And this is the great deception of sense." For a recent paper emphasizing the centrality of accidents to perception in Hobbes, see Callaghan, 2001.

[17] *Strictu sensu*, it is a function of forgetting: the dilution of a sense impression, which occurs because we turn our attention to new impressions, is memory. It is this dilution that causes me to retain some attributes and not others.

[18] He refers to the medieval "suppositum" on the following page as he applies this theory to the Trinity.

[19] Hobbes's insistence on homogeneity in his mathematical writings also indirectly suggests the view of accidents I am attributing to him: a discursive matrix presupposes agreement on which set of real properties *are to be taken* as relevant in communication and counting. In counting red things, they are all different, both in the nature of their redness and having indefinitely many other real properties, *but* those differences do not matter. That nonetheless there is some (arbitrary) relation between the discursive field and the objects in question is why propositions about them (= natural science) are contingently true. The absence of real moral properties is why propositions about them can be necessarily true.

[20] For a discussion of Hobbes's rejection of intelligible species, emphasizing how his account of the origination of perception differs from the medievals, Lee, 2000, pp. 34–45; Leijenhorst, 2001, pp. 56–100 and 2007, pp. 84–94.

[21] A possible source for this position would be the Lullist guides to memory technique, which relied on "commonplaces": imagined locations (e.g., rooms in a house) which were to associatively bring to mind whatever one was to remember. Hobbes would thus be taking the de facto position that all memory is, in this sense, "artificial." For the memory manuals, see Rossi, 2000.

[22] AT VII, 172; PWD II, 122; I have revised the translation along the lines suggested in Ross 1988, p. 223. On this exchange, see Zarka, 1987, *passim*; Ross, 1988; and Sepper, 1988. Consideration of this text does not require settling the debate about the extent to which Hobbes is responding to Descartes more generally; for the view that he is, see Tuck, 1988; for the other side, see Leijenhorst, 2001, pp. 11–15.

[23] AT VII, 178; PWD II, 125–6.

[24] Either that, or he is moved to atheism. Now is not the place to rehearse this interpretive debate. I have been inclined to take Hobbes's religious pronouncements as sincere: see Hull, 2002. For some of the background debate, see Martinich, 2001 and Rogers, 1990 (both interpreting Hobbes as a theist); and Curley, 1996 and Jesseph, 2002 (arguing that Hobbes is an atheist). Forster, 2003, offers a detailed discussion of the Curley-Martinich debate, siding with Curley, and adds the argument that Hobbes fails to authorize the Bible: he offers no reason why the Bible should be accepted as an authority, and appeals to faith on the unicity of the sovereign's interpretive power. I will add only three points here: (a) Hobbes offers a cosmological (first cause) proof *that* God exists L 11.25, 74–5 (the passage also alludes to an argument by design), and distinguishes this path to belief in God from the fear that motivated the pagans (L 12.6, 76–7). Since this is an argument from experience, and since arguments from experience cannot attain certainty (see my discussion of the *regressus* problem), one could ask little more from natural reason (for skepticism about the sincerity of this proof—and whether it coheres with *De Corpore*—see Jesseph, 2002, pp. 150–2). For more on Hobbes's God as demonstrable but unknowable, see Pacchi, 1988 and Zarka, 1984, pp. 169–70; for nominalism as a piety in Hobbesian terms, see Roux, 1990. Limiting the role of natural reason, and increasing the role of faith, was a

standard tactic of nominalist writers. Ockham, in particular, seems to have done precisely this in response to the condemnation of Aristotelian elements in Thomism (on this, see Lee, 2002). In other words, that a nominalist who was critiquing Aristotle might emphasize divine power and relegate other aspects of our knowledge of God to faith is not particularly surprising (in any event, differences in faith can be substantial, even among those who turn out to agree on most points of natural reason: for a forceful demonstration of this point in the case of Maimonides and Aquinas, see Dobbs-Weinstein, 1995). (b) Forster cites Hobbes's discussion of the trinity (which he ratchets to his new theory of personhood (L 16.12, 114)) as obviously heterodox. The difficulty here, which Forster shows in detail, is that many of the church orthodoxies depended on Aristotelian terms. It was thus extremely difficult to reconcile a new, anti-Aristotelian physics with the demands of orthodoxy. Hobbes's difficulties here should be paralleled with Descartes' tortured attempts to explain to Arnauld the orthodoxy of his physics, given its obvious difficulty in explaining transubstantiation in the (Aristotelian) terms mandated by the Council of Trent. One should recall that this difficulty is what ultimately led to the censure of Cartesianism (on this, see especially Ariew,1999 and Schmaltz, 2002). (c) Whether or not Hobbes believes in God is ultimately irrelevant to the political question of the regulation of faith and the reading of scripture. As I will indicate in my discussion of the state of nature, religious belief is a sociological fact for Hobbes, whether or not it is true (this point is also emphasized in the closing pages of Jesseph, 2002).

[25] On this argument, see, for example, Zarka: "words don't possess any signification by nature, and it would therefore be totally in vain to search in current language for the traces of a forgotten natural language" (1987, p. 86). Three obvious targets of these remarks are: scholastic realism, the belief in the semantic anchoring function of Biblical Hebrew, and the belief in some sort of Caballistic or Lullist ordering schema behind language. All were prevalent in the seventeenth century. For Biblical Hebrew, see Kottman, 1975. For the Lullism, see Rossi, 2000. Hanson suggests: "thoroughgoing conventionalism . . . pulls the linchpin of the magical world of the renaissance" (1991, p. 631; see p. 642 for political implications). As the following will indicate, while I am generally sympathetic to Hanson's account, I do not think that Hobbes "assumes that there is a kind of original innocence" of signification (1991, pp. 645–6; he elsewhere claims that mental discourse represents "thought uncontaminated by the deceptions that speech makes possible," 1993, p. 648). Rather, the state of nature allows political science to produce such innocence by fiat.

[26] The Latin text tempers somewhat what might sound like an atheistic implication of this denial: "for *finis ultimus* and *summum bonum*, of which ancient ethicists speak, have no place in the present life [Finis enim ultimus et summum bonum, de quibus loquuntur ethici veteres, locum in praesente vita nullum habent]" (OL III, 77).

[27] On this point, see also Miner, 2001; Watkins, 1965, pp. 150–7; Zarka, 1987; and Zarka, 1995, esp. pp. 65–126.

Chapter 5

[1] For these changes, see Tricaud, 1988.

[2] This subtractive process is emphasized in Pasquino, 1994; on the negative imagery, see also Chanteur, 1969.

[3] See Skinner, 1996, esp. pp. 11–133.

[4] I take this point to be definitively established in Skinner, 1996. Other texts that explore the dual purpose of *Leviathan*—polemical and scientific—include Ashcraft, 1971 (exploring especially the resonances of *Leviathan*'s description with contemporary accounts of Native Americans); Farneti, 2001 (analyzing Hobbes's use of the Leviathan as an effort at a counter-mythology), Hull, 2002 (analyzing Hobbes's use of *Job* in light of seventeenth-century commentaries on the text); Malcolm, 2002g (on the possible referents for the title page image); Patapan, 2000 (on Hobbes's rhetorical effort to unsettle pride); Skinner, 1972 (on *Leviathan* as an intervention in the engagement controversy); and Tarlton, 1996 (showing that *Leviathan* can be read both as a theoretical text and as an advice manual for princes). One reason that Hobbes would be so careful to use good rhetoric is that it is entailed by his theory, which emphasizes that political objectives are achieved through the control of language and signification. This point is emphasized in Ball, 1995; and Zarka, 1995.

[5] This point is emphasized in Pasquino, 1994, pp. 200–1. For the increasing importance of images to seventeenth-century discourse generally, see Burke, 2003. There are numerous other possible resonances between Hobbes's imagery and both classical and Biblical sources; for a recent discussion, see Hoekstra, 2007.

The exegetical piety that "Hobbes contradicts himself because he both uses metaphor and denounces it" has long outlived its usefulness. As Hobbes reminds his readers, "reason and eloquence (though not perhaps in the natural sciences, yet in the moral) may stand very well together. For wheresoever there is place for adorning and preferring of error there is much more place for adorning and preferring of truth, if they have it to adorn" (L R&C 4, 484). Skinner, 1996, which in this respect builds on Johnston, 1986, remains the most comprehensive study. See also Stillman, 1995 (arguing that Hobbes's "writing is committed to perform for us (and ultimately . . . to have us perform) a transformation of metaphor into logic—a transformation, it should be stressed from the start, that aims at a strangely magical event, the incarnation of sovereign power," pp. 799–800); and Wilson-Quayle, 1996 (showing that Hobbes approved of enlightening images, even as he disapproved of deceptive metaphors).

[6] I thus think the question of the "natural emergence" of this condition, as pursued, for example, in Gauthier, 1988, is misplaced. Gauthier seems to me to be more correct when he says Hobbes's argument proceeds by "showing that the absence of any social structure—the state of nature—would be intolerable, and that only absolute sovereignty truly safeguards against that absence" (1987, p. 294). Watkins suggests the state of nature "is an 'ideal' or limiting case in which every vestige of authority and organization has been imagined away" (1965, p. 47). See also Goyard-Fabre, 1995, pp. 82, 88 (emphasizing the methodological continuity between the *annihilatio mundi* of the *Elements of Law*). From all of this it also follows that—no matter how frequently pursued—questions about how individuals might get themselves out of the state of nature are also only relevant insofar as they illustrate the logic of his position. On this point, see, for example,

Johnson, 1982 ("Hobbes need no more provide a solution to how men in the state of nature are to get out of it, than a physicist must provide us with a way to get a body moving in a pure inertial state to curve," p. 42); Pasquino, 2001 ("the question is not to try to figure out how to escape it, but simply to understand that the concept is used rhetorically, in order to change the mind of those involved in religious civil war," p. 408); and Tricaud, 1988 ("it is a *model* (taking the word in such sense as physicists and economists make use of), whose function is not to *reproduce* the true condition of mankind, but to *illuminate* it. . . . Even if 'nature' is thought of in contradistinction to 'art', the notion of a state of nature is a conceptual artefact," pp. 110–11).

7 It is this feature of Thucydides' account that leads contemporary writers to regard it as the first "modern" history. On the complicated linkages between this "linear causal perspective" and Hobbesian thought, see Brown, 1989. For the general cultural and literary background to Hobbes's work in this regard, see Schlatter, 1945.

8 *De Cive* and *Elements of Law* are less precise. *De Cive* reports: "the present century presents an example [*exemplum . . . exhibet*] of this in the Americans. Past centuries show us nations, now civilized and flourishing, whose inhabitants then were few, savage, short-lived and mean, and lacked all the comforts and amenities of life which *peace* and society afford" (DC I.13, emphasis in original). *Elements of Law* reports on "the experience of savage nations that live at this day, and by the histories of our ancestors, the old inhabitants of Germany and other now civil countries" (EL 14.12).

9 For more on this, see Malcolm, 2002f.

10 As Curley points out, the Biblical reference does not seem apt, since God immediately punishes Cain. See Curley's note to the passage in his edition of *Leviathan* (Hobbes, 1994c, p. 77, n7). Pasquino suggests that the example would have been quite powerful to a seventeenth-century reader, and might have been intended to show the limitations of paternal and divine power as deterrents (1994, p. 300).

11 For a somewhat different analysis of the sedition and subsequent stasis along Hobbesian lines, see Brown, 1987.

12 Thucydides writes, "the received value of names imposed for signification of things, was changed into arbitrary." (III.82; EW VIII, 348). Thucydides is talking specifically about moral words: for example, "inconsiderate boldness" becomes "true-hearted manliness" and "modesty, the cloak of cowardice." Terence Ball emphasizes the continuity of Hobbes's emphasis on the control of signification from the Thucydides translation onwards; see Ball, 1987, 1995; see also Reik, 1977, pp. 36–52. For the Renaissance, see Reiss, 1992, 42-69.

13 Cited in Ashcraft, 1971, p. 1094; Ashcraft concludes that "while the Indians were almost universally regarded by Englishmen as culturally inferior beings, they were not generally viewed as collections of politically hostile tribes" (1095).

14 The Latin also indicates more clearly the continuity between this form of rule and others by emphasizing the harmony of affects: the concord of "natural lust"

in the English becomes "similarity of desire [*similitudine cupiditatum*]" in the Latin. More generally, Hobbes's break with the English doxa on Native Americans suggests that it would be worthwhile to consider his exposure to Spanish debates. In 1550–51, indigenous rights had been officially debated in Spain at Valladolid, with Juan Gines de Sepulveda arguing in favor of the total and immediate subjection of the indigenous Americans and Bartolome de las Casas arguing against (he favored a program of persuasive conversion). Although we do not know the extent of Hobbes's familiarity with these debates, it seems reasonable to suppose he would have at least some knowledge of them. On the one hand, he otherwise evidences familiarity with sixteenth-century Spanish work in figures such as Suárez. On the other hand, Grotius had adopted some of the more liberal Spanish views—particularly the idea that the Americans were not legally empty—in his *Mare Liberum* (1609). For a summary of the Spanish debates, see Davidson, 1994. Hobbes also had a documentable familiarity with the administration of colonial Virginia by way of his employment in the Cavendish household: see Malcolm, 2002e.

15 My discussion on this point is heavily indebted to Pateman, 1988, esp. at pp. 48–9. In this regard, Hobbes breaks with almost everyone else writing in the seventeenth-century. Ashcraft's remark that "Hobbes simply incorporated the theory of patriarchism into his description of the state of nature" (1971, p. 1106) thus requires some qualification. By nature, primogeniture is denigrated to the conceptual equivalence of first seizure, and is only for things which can be neither divided nor enjoyed in common, and for which there is no agreed allocation (L 15.27, 108).

16 The parallels between this passage and the *Leviathan* passage are analyzed in Klosko and Rice, 1985. Hobbes also picks up the discussion of invasions of agriculture at L 13.3, 87.

17 I do not claim this is the only such theoretical reason. For example, Carole Pateman argues that the state of nature could not subsist and therefore could not be the origin of a society: it would be irrational to weaken oneself by taking care of a child; hence, "all stories of original social contracts and civil society are nonsense because the individuals in the state of nature would be the last generation" (1988, p. 49).

18 EW VII, 174; see the discussion of this passage in chapter 2.2, "Thinking as Construction."

19 For further commentary on this passage, see, for example, Sorell, 2007.

20 Separating Hobbes from Bacon, he cajoles that "for as my *Lord Bacon* wisely observes, nothing has more hindered the growth of Learning than peoples studying of *new words,* and spending their time in *chaptering, modelling,* and *marshalling* of Sciences" (Eachard, 1958, p. 21).

21 For other passages, see the discussion in ch. 2.3.

22 Hoekstra (1997) makes a persuasive case that the fool in question here does not just think but advocates injustice. Without denying the importance of this point, I think the embeddedness of the discussion in Carneades (the classical discussion does not seem to clearly distinguish between silent and explicit fools) indicates

that more can be said. In particular, the point I want to press here is that the question of promise-breaking is fundamentally different in and out of a commonwealth. Hoekstra, 1999, 232–3, emphasizes that Hobbes *also* provides reasons for the silent fool to reconsider his position, even if that is not the main point.

[23] See Skinner, 1996, pp. 9–10. In addition to Cicero, a classical source for this attribution is Quintilian: "After all, the Academics argue both sides of a question, but live according to one side only, and the great Carneades, who is said to have spoken at Rome in the presence of the censor Cato just as vigorously against justice as he had spoken in defence of justice the day before, was a perfectly just man" (*Institutio Oratoria* 12.1.35).

[24] For the Locke, see his *Essays on the Laws of Nature* VIII (in 1997, pp. 127–33). Matheron suggests that Hobbes essentially adopts Grotius' understanding of natural law, but without the natural instinct for sociability. For Hobbes, then, we not only have a right to preserve ourselves, we lack the right not to preserve ourselves (1986a, p. 87). Matheron elsewhere suggests that Hobbes's natural law might be seen (against Machiavelli) as an inverted Thomism: "de ce thomisme inverse, la normativité subsiste. Il y a une *loi naturelle*, qui nous *oblige* à faire ce qui est indispensable à notre conservation et qui nous *interdit* de faire le contraire" (1986b, p. 77, emphasis in original). I take it that my emphasis on *poietic* elements (as opposed to *praxis*) in Hobbes distinguishes the reading here from Matheron's, in that, to the extent that Hobbes's thought is *poietic*, the ends of reason tend to separate from their natural referents. In other words, the *potentia/potestas* relation in Hobbes is both difficult and ambiguous. See note 28, below; and ch. 7, note 5, and my remarks about immanence and transcendence in that chapter.

[25] There are also lost pages of Cicero which referred to the same discussion: see *De Re Publica* III, 8–9. A line from Cicero's *De Legibus* might serve as a leitmotiv to understand Hobbes's engagement with skepticism: "And let us implore the Academy—the new one, formed by Arcesilaus and Carneades—to be silent, since it contributes nothing but confusion to all these problems. . . . At the same time I should like to win over this school, and so do not dare to banish it from the discussion" (*De Legibus* I, 39).

[26] In other words, both first and second performance become rational, and at the same time. On this point, I agree with Gauthier's (1988) description of the contract as "only partially coordinative"—that is, though everyone involved prefers compliance, no one expects it without the addition of sanctions and incentives whose only purpose is inducing compliance. Gauthier offers this account in opposition to Hampton, 1986, which characterizes the initial move into sovereignty as a coordinative agreement where compliance is motivated only by the terms local to that agreement. Gauthier elsewhere suggests that the hypothetical pareto-maximizers might find it rational to change their concept of rationality to one that allows contracts (1977, pp. 156–7).

[27] Eternal damnation does make an appearance later: "as for the Instance of gaining the secure and perpetual felicity of Heaven, by any way; it is frivolous: there being but one way imaginable, and that is not breaking, but keeping of covenant" (L 15.6, 103). G. A. J. Rogers interprets this passage as evidence of Hobbes's theistic sincerity. It strikes me as conditional: if someone thinks that revelation will get her to heaven, then . . . Rogers suggests that this line of reasoning will "only

weigh with theists" (1990, p. 48)—but only a theist would take the position it is designed to refute.

28 Such a move is only to be expected, given Hobbesian theses about the unknowability of God. Hobbes is in any case often credited with the move. Thus "humanizing the notion of right, which according to tradition had always been rooted in divine transcendence" achieves a "doctrinal revolution" (1988, p. 18 and *passim*). On this, see also Balibar, 1996 and Zarka, 1999b. As the following discussion should indicate, Hobbes's views on the reduction of right to power are complicated, and I would hesitate to read him as pushing the point as far as, say, Spinoza. I take it that the desire to ground the Hobbesian laws of nature in God is what drives the so-called Taylor-Warrender thesis, which as the discussion here should indicate, I think is clearly a misreading. For an early critique, see Watkins, 1965, pp. 59–61. For a survey of seventeenth-century efforts to deal with this problem, see Zarka, 1999a.

29 Cf. "Before the names of Just, and Unjust can have place, there must be some coërcive Power, to compell men equally to the performance of their Covenants" (L 15.3, 100–1). Hoekstra suggests that Hobbes's alignment of justice with reason makes his discussion of the possibility of justice in the state of nature inconsistent (2007, p. 120); it seems to me that there is no inconsistency in reasonably desiring justice but also reasonably not taking the risk of being a first performer.

30 The Latin is more also more explicit that coercive power is both necessary and sufficient for meaningful contracts: "for the question is not of mutual promises in the natural condition of man, where there is no coercive power, for those promises are not pacts [*quaestio enim non est de promissis mutuis in conditione hominum naturali, ubi nulla est potentia cogens; nam sic promissa illa pacta non essent*]" (OL III, 113).

31 One should recall that for Hobbes "will" is the name of the last affect before an action. Interpreting someone's will is thus the same process by which one would interpret any other activity on their part; see, for example, *Hobbes and Bramhall* §§25, 32, and 34; and in *Leviathan* at L 6, 44–5; and L 46, 468.

32 This point—that the conflict does not have to be actual fighting—is emphasized in Curley, 1989, p. 176; and by Foucault: "there are no battles in Hobbes's primitive war, there is no blood and there are no corpses. There are presentations, manifestations, signs, emphatic expressions, wiles, and deceitful expressions; there are traps, intentions disguised as their opposite, and worries disguised as certainties" (2003, p. 92). Hardt and Negri draw the link to the current global situation, suggesting that "in this context, war has become a general condition. . . . That erodes the distinction between war and peace such that we can no longer imagine or even hope for a real peace" (2004, pp. 4–5).

33 See, for example, Kavka, 1983, arguing that they can, and Hampton, 1985, arguing that they cannot. One feature of many accounts that enable actors in the state of nature to emerge from it through voluntary associations is an assumption of communicative transparency. Kavka, for example, suggests that a "gap" in Hobbes's argument is that "for a universal state of war to exist, it is not enough that anticipation be the most reasonable strategy, it must also be *believed* so by all who do not wish to fight for other reasons, and most must be aware that others so believe it. . . . For only then is it guaranteed that 'the will of each to contend by

battle is sufficiently known'" (1983, p. 58, citing L 13.8, 88). My argument here is that the *scientia* requirement is impossible to meet, which means that it will be rational to assume that others are out to get you. The Latin text drops the sufficiency requirement: "consistit enim natura belli, non in pugna, sed in tractu aliquo temporis, quo durante voluntas armis decertandi est manifesta" (OL III, 99). The question is what it takes for a "sign" to be "manifest" in the state of nature, and the appropriate strategy for dealing with those conditions. Cf. Pasquino, referring to the "condition of epistemic opacity which destroys the preconditions for any rational trust" (2001, p. 407). He cites *Elements of Law*: "they must needs provoke one another by words, and other signs of contempt and hatred, which are incident to all comparison" (EL 14.4). Cf. also Zarka, remarking that "in the state of nature, individuals are not able to distinguish with certainty that which is necessary to their preservation from that which is not, or even from that which is contrary to it" (1995, p. 200). This is partly because "in this condition of general insecurity, where all the signs of human relations are equivocal, the concern with self-defense leads each individual to anticipate the aggression—real or imaginary—of which he could at every moment be the object" (1995, p. 199). He also suggests that the desire for glory—for signs from others which affirm one's own power—is what makes war in the state of nature irrational (1995, p. 145).

34 I am using the term "semiotic" specifically to indicate the failure of language as a system of signs, or, in the more historically precise sense developed in the previous chapter, as a system of "signification." In the chapter on speech, Hobbes indicates that this is one of two primary uses of speech (the other is as a mnemonic device): "when many use the same words, to signifie (by their connexion and order,) one to another, what they conceive, or think of each matter; and also what they desire, feare, or have any other passion for" (L 4.3, 25; this is repeated almost verbatim in the Latin at OL III, 22–3, using the technical "*significare*").

35 Hampton (1985) argues that the "passions account" is insufficient to generate the state of nature. I do not think that anything here hangs on whether or not the various factors are individually sufficient to cause the state of nature conflict, and whether Hobbes has thereby overdetermined the *bellum omnium*. I take it that they are clearly jointly sufficient to cause it. Hampton elsewhere (1991) emphasizes the lack of information individuals have in the state of nature as one of the factors which makes emergence from it an insoluble problem; in this I think she is correct. See also Vanden Houten (2002) for the thought that passions can undermine prudential reasoning, despite apparent Hobbesian assurances to the contrary. Short-sightedness, which Hampton takes to be essential to a sufficient explanation of the *bellum omnium*, is analyzed by Vanden Houten (270) as a failure of prudence. Slomp (2007) suggests that glory is the passion most implicated in state of nature conflicts in the sense that it is necessary for any conflict to get started.

36 See the important discussion in Slomp, 2007.

37 For further discussion of the *annihilatio mundi*, see Lee, 2004, pp. 112–13; Leijenhorst, 2001, pp. 53–5; and Zarka, 1987, pp. 36–58. Zarka emphasizes the Hobbesian break with both Ockham and Descartes: "the sign loses the natural anchorage in things which characterizes it in Ockham. This possibility of

studying representation independently of things is attested in Hobbes by the persistence of representation in the soul when the cause which produces it is no longer present, or, according to the annihilatory hypothesis, when it is destroyed" (1987, p. 52).

38 As these passages suggest, it is the Hobbesian elevation of the affects and imagination (discussed in the previous chapter) that enable this conclusion. Consequently, the tendency to read Hobbes as an "individualist" or an "egoist" (depending on how one takes that term) is misplaced. For Hobbes, the individual is not an atomic unit of analysis. Zarka puts the point provocatively: "thus the *ego*, an object of knowledge confused by sentiment, is not part of the truths established by the *philosophia prima* of Hobbes, and it never accedes to the status of a first truth. . . . Moreover, it is because neither the thing nor the ego has the status of a foundation that a foundation for politics becomes possible" (1987, p. 43, emphasis in original).

39 Cf.: "And because the constitution of a mans Body, is in continuall mutation; it is impossible that all the same things should alwayes cause in him the same Appetites, and Aversions: much lesse can all men consent, in the Desire of almost any one and the same Object" (L 6.6, 39).

40 This point is discussed as an aspect of Hobbes's engagement with skepticism in Missner, 1983. Cf. also Slomp: "more than ever before in *Leviathan* Hobbes suggests that the motivation of people is varied, changing and unknown. In *Leviathan* no unifying principle of motivation replaces glory, but the pursuit of power allows us to explain and predict the behavior of individuals regardless of their motivation" (2007, pp. 186–7).

41 "In the condition of meer Nature, the inequality of Power is not discerned, but by the event of Battell" (L 14.31, 99).

42 Cf. also L 11.26–7, 75, which makes the same argument, including the point about manipulation. See also the account in chapter 12, which underscores that anxiety about the future is a necessary part of being human (12.5, 76). In this sense, the fearful are vulnerable to Cartesian doubt about whether a *genie malin* is manipulating them. For the suggestion of a parallel between the Cartesian evil demon and a malevolent Machiavellian prince, see Richir, 1997. For a discussion of the roots of Hobbesian fear in human ignorance, see Blits, 1989.

43 The passage is perhaps clearer in the Latin—the one who fears first, sees some cause of the fear ("qui primus metuit, metuendi causam vidit aliquam" (OL III, 45). Cf. L 11.16, 73, where Hobbes notes that ignorance of causes forces one to rely on the "advise, and authority of others."

44 On this see also Hanson, 1993, p. 654, tracing how for Hobbes the common people's prudential judgment is easily manipulated by elites.

45 This is not to pass judgment on Hobbes's economic views. Clearly McPherson's thesis (1962) requires revision, as has been noted by numerous commentators (though for a more careful analysis, showing that many of the critiques of McPherson misread him, see Townshend, 1999). McPherson's claim that the state of nature is ideological (the state of nature "is not about 'natural' man as opposed to civilized man but it is about men whose desires are specifically civilized . . . the state of nature is the hypothetical condition in which men as they now are, with natures formed by living in civilized society, would necessarily find themselves if

there were no common power able to overawe them all" (18–19)) seems correct but not particularly helpful in locating Hobbes, since the thought that Hobbesian people desire to "live well" (24) does not differentiate them from Aristotelian people. A discussion of Hobbesian economic thought, or Hobbes's importance to later economic thought, should probably begin with his friend William Petty.

46 Many will be afflicted with diffidence, but of course to display such diffidence is to invite conquest, and so every one will be concerned to project strength. When others undervalue that strength, conflict ensues.

47 Thus, vainglory is infectious, as romance stories generate the desire to emulate: see Kahn, 2001. Tricaud suggests that Hobbes's model on this point was the widespread seventeenth-century practice of dueling (1982, pp. 118–19).

48 Pasquino, 2001, n17 *et passim*. Against the idea that the state of nature could be an iterative prisoner's dilemma (favored by Hampton 1986), see also Hoekstra: "iteration itself will not be reliable, for the dilemma may never recur with the same people" (2007, p. 115).

49 For a thorough discussion of deterrence outcomes in a crisis situation characterized by imperfect information, see Kilgour and Zagare, 1991.

50 This point is not emphasized by Hobbes. I do think it follows from his discussion; I draw it from work in deterrence theory. See, for example, Blair, 1993.

51 "Experience" here implies the Aristotelian "way things normally happen." Hobbes is not inviting the reader to attempt an experiment in the eighteenth-century sense. On the difference between the two, see Dear, 1995 and Milton, 1987 (noting that Hobbes is openly contemptuous of induction by partial enumeration (p. 61, citing OL IV, 179)). See also ch. 3, n11 and accompanying text.

Chapter 6

1 This negativity is particularly emphasized in Chanteur, 1969.

2 Deborah Baumgold (2004) argues that this chapter's text—especially the claim that democracy is the first form of government (EL 21.1–2), which Hobbes later drops—suggests that it was composed at a later date than others, and in a greater hurry. The discussion of the "multitude," however, seems clearly present in Hobbes's later work, if not in the exact form it is found here.

3 On this, see Tuck, 1993, 42ff; see also Ashcraft, 1986, 161 n134 and 301ff; for other English condemnations contemporary to Hobbes.

4 Quoted in Ashcraft, 1986, 161 n134. Ashcraft remarks here that "the basic framework for every critique of political radicalism in the seventeenth century, from Hobbes to Samuel Parker, can be found in Ireton's speeches at Putney" (*ibid.*).

5 Quoted in Ashcraft, 1986, pp. 301–2. The latter two lines are from Samuel Parker's *Discourse on Ecclesiastical Polity* and *Defence and Continuation of the Ecclesiastical Polity*.

6 These are the theses of Ashcraft (1986) and Pateman (1988), respectively. Locke does seem to use the term polemically, and against Hobbes, at least once. When answering the objection that people are bad judges in their own cases, he con-

cedes the civil government is the only proper remedy, and turns the argument into a critique of monarchy: "I desire to know what kind of Government that is, and how much better it is than the State of Nature, where one Man commanding a multitude, has the Liberty to be Judge in his own Case" (2T, 13). I discuss this passage in detail in the closing section on judicature.

7 I draw these examples, which are presented in the form of an objection, from Copp, 1980, p. 604.

8 Compare Foucault on the Renaissance: "Scriptural commentary, commentaries on Ancient authors, commentaries on the accounts of travelers, commentaries on legends and fables: none of these forms of discourse is required to justify its [the form of discourse's] claim to be expressing a truth before it is interpreted; all that is required of it is the possibility of talking about it. Language contains its own inner principle of proliferation" (1970, p. 40). Like our own contemporary political scientists, Hobbes is concerned to stop proliferation at any cost. Foucault earlier points directly to the problem: "sixteenth century discourse condemned itself to never knowing anything but the same thing, and to knowing that thing only at the unattainable end of an endless journey" (30). Hobbes specifically references scholastic theologies, which "once fallen into this Error of *Separated Essences,* they are thereby necessarily involved in many other absurdities that follow it" (L 46.19, 466), adding "one Inconvenience into another, without end, and without number" (L 46.23, 467). For commentary on these passages, see Skinner, 1996, pp. 397–8.

9 As should already be apparent, Hobbes is following tradition in assuming that, as Reiss puts it, "the idea of a union between language and the social real was an a priori" (1992, p. 53; cf. generally pp. 42–69).

10 The beginning of the paragraph verifies the causal relationship: "But it is an easy thing, for men to be deceived, by the specious name of Libertie; and for want of Judgement to distinguish, mistake that for their Private Inheritance, and Birth right, which is the right of the Publique only. And when the same error is confirmed by the authority of men in reputation for their writings in this subject, it is no wonder if it produce sedition, and change of Government" (L 21, 149). For a thorough analysis of Hobbes's usage of "liberty," see Skinner, 1990.

11 See Deleuze, 1994; specifically, Deleuze defines Platonism as follows: "the true Platonic distinction . . . [is] not between the original and the image but between two kinds of images [*idoles*], of which copies [*icônes*] are only the first kind, the other being simulacra [*phantasmes*]. The model-copy distinction is there only in order to found and apply the copy-simulacra distinction. . . . The function of the notion of the model is not to oppose the world of images in its entirety but to select the good images, the icons which resemble from within, and eliminate the bad images or simulacra" (127). Also, "Ideas inaugurate or ground the world of representation. As for the rebellious images which lack resemblance [*simulacra*], these are eliminated, rejected and denounced as ungrounded, false claimants" (p. 272; original text terms provided with the translation).

[12] Cf. Borrelli (1997), tracing the problem of political prudence and obedience from Bodin to its resolution in legal institutions by Hobbes. As he puts it, "the civil law becomes a juridical system productive of obedience, a positive instrument of discipline" (23). Hobbes's legal thought has been neglected in recent scholarship, and his judging has been neglected within that. Boyle (1986) speculates that this may in part be because, like the legal positivists, Hobbes insists on definition, but that, unlike the positivists, Hobbes rejects any notion of the autonomy of the law. The latter point—that Hobbes views legal theory not as something autonomous but as an integral part of political theory—is central to what follows, and Boyle makes the point quite forcefully.

[13] That these are technical terms, specific to Hobbes's discussion, is evident in his repeated use of "cives" in the Latin version: "Lex civilis unicuique civi est regula qua civitas . . . ad distinctionem boni et mali ute imperat" (OL III, 197). I return to the sense of law as "regula" or "mensura," below.

[14] As Zarka puts it, "contrary to the caricatured image which has often been given, the practical efficacy of political philosophy rests in Hobbes on the idea of the usage and development of reason, on the side of the holder of power as well as that of the subjects" (1995, p. 47). He says later that "there therefore ought to be a self-foundation of law and of the fact of the social convention; in effect it ought itself to create the conditions of its proper juridical validity and of its proper effectivity. In other words, the social convention will become that which is not able to be contested either in law or in fact" (212).

[15] Hobbes was perhaps generous in his attribution of consensus; as Carter (1979) points out, Hobbes's England was both extremely litigious and presented a veritable state of nature among understandings of the law. On this, see also Jones, 1971, pp. 32–52.

[16] Thus in the Review and Conclusion, Hobbes marks as a possible defect of commonwealths, "their Imperfect Generation, consisting in the want of an Absolute and Arbitrary Legislative Power; for want whereof the Civill Soveraign is fain to handle the Sword of Justice inconstantly" (L R&C 8, 486). "Arbitrary" here thus means "at will," but it is precisely the subsumption of "arbitrary" into "absolute" that rules out the power being "arbitrary" in the sense of "random." This point is noticed by recent scholarship which emphasizes the "bearer" of the Hobbesian sovereignty, rather than the sovereign. As Ludwig puts it, "The sovereign 'carries' the *person* of the commonwealth as perhaps the round body carries the *figure* of the circle" (1995, p. 59, emphasis in original). This distinction is possible because both the sovereign and the commonwealth are "artificial" persons. For the complexities of Hobbes on artificial persons, see Copp (1980), and then the debate between David Runciman and Quentin Skinner: Runciman, 1997, pp. 6–33, and 2000; and Skinner, 1999. Goyard-Fabre emphasizes the connections between Hobbes's understanding of artificial personhood and his attack on older natural law traditions (1982, esp. pp. 26–7). Tricaud argues that the English *Leviathan* differs significantly from the Latin on this point (as does Skinner, though they

offer divergent accounts of this difference), suggesting that in the English edition, "Hobbes is obviously anxious not to allow any kind of corporate existence of the subjects outside the person of the sovereign" (1982, p. 95). This point should be retained, with the caveat that "subjects" are co-constituted with the commonwealth, as opposed to the "multitude." See ch. 6.1 "From Multitude to People," above.

[17] For a recent version, see Tarlton (2002a), arguing that Hobbes should be understood as advocating "despotism," that is, "for Hobbes, viable government always and necessarily constituted a system of absolute, arbitrary and unlimited political power" (60). Tarlton (2001) complains that the despotism critique has become less common in more recent Hobbes interpretation.

[18] This is perhaps somewhat of a caricature of Kant's position; in articulating it, he does suggest that "this problem is both the most difficult and the last to be solved by the human race" (*ibid.*), and when read in combination with his "What is Enlightenment" essay, the "History" essay does offer at least a skeletal account of the emergence of a "rational" society.

[19] For the distinction, and an argument that politics itself is founded on the incommensurability between these two orders—one "arithmetic," of shopkeepers; and the other "geometric," of political philosophy—see Rancière, 1999, esp. pp. 15–19.

[20] Cf. Lloyd (2001), pointing out the connection between equity and submission of private disputes to public arbitration: I am to defer to public judgment because I would not think it reasonable for others to exercise their private judgment in matters of common concern. See also Polin, suggesting that "the laws of nature are nothing other than the explication of this rule—the golden rule," which Hobbes calls "one easie sum, intelligible even to the meanest capacity" (L 15.35, 109)—"in the principle circumstances of life in society" (1995, p. 59).

[21] For a thorough analysis of Hobbes's use of the term, see Hoekstra, 2001. On the absence of discussion of tyranny, see also Bertman, 1990.

[22] The Latin reaches the same result somewhat more directly; cf. OL III, 137.

[23] Hobbes invokes this tradition in *De Corpore*: a concrete term is "quod rei alicujus, quae existere supponitur, nomen est, ideoque quandoque *suppositum*, quandoque *subjectum*, Graece *hupokeimenon* appellatur ut *corpus* [etc.]" (DCo III.2, emphasis in original). For an illuminating discussion of some difficulties in Ockham's view, see Normore (1997). Normore notes that Ockham ends up relying on the will of the user of terms to disambiguate the various references in things supposited; in this, Ockham might be seen to be a precursor to Hobbes.

[24] Clearly much of rides on how one is to understand "authorization" here. Skinner suggests that to be "authorized" is "to have been granted a commission or license to perform an action by some person or persons who must possess the right to perform it themselves. The grant must take the form of a voluntary transfer of right, since commissioning and licensing are names of voluntary acts." (1999, p. 9). The bearer of the sovereignty thus, as part of executing his commission as sovereign, thus appoints a further proxy to deal with questions of legal interpretation.

[25] For discussion and further references, see Schuhmann (2004c), to which the following discussion is considerably indebted, especially for the references to scholastic sources. Schuhmann points out that the Hobbesian definition of law repeats almost verbatim that found in Suárez, *De Legibus* I 1, 2–7.

[26] "Lex quaedam regula est et mensura actuum secundum quam inducitur aliquis ad agendum, vel ab agendo retrahitur" (STh I–II q 90 a 1 co.). The passage is cited in Schuhmann, 2004c, p. 177. Thomas discusses the necessity of promulgation at STh I–II q 90 a 4.

[27] One should therefore ask questions about the violence of imposing the law in the first place, regardless of its results. Hobbes criticizes those who require mythologization of the past, noting that "there is scarce a commonwealth in the world whose beginnings can in conscience be justified" (L R&C, 8, 486). See also Cover (1986), noting that "we must expect . . . to find in the act of interpretation an attention to the *conditions of effective domination*" (1616, emphasis in original); and Zarka: "Hobbes perceives, perhaps more clearly than others, the paradoxical character of politics, always stretched between language and violence, law and power, reason and passions" (1995, p. 20).

[28] In this respect, equity functions similarly to its role Plato's *Statesman* and Aristotle's *Ethics*, as a "correction of what is legally just" (*Ethics* 1137b12). It is important to note that for Hobbes, equity is what mediates between the risks of what would now be called originalism and unrestrained judicial activism.

Chapter 7

[1] Thus the central roles played by Coase (1960) and Calabresi and Melamed (1972). For early and substantial critiques of this movement, see, for example, Baker, 1975 and Schlag, 1989.

[2] For a meditation on these points, see especially Waldron, 2002.

[3] Foucault reads Hobbes as attempting to contest any effort to legitimize the paternity of a ruler (king or otherwise) through appeals to inherited rights of conquest (2003, pp. 98–110).

[4] "quam rem unam in hoc libro non demonstratam sed probabiliter positam esse confiteor" (DC Pfc. 22). Jean Hampton thinks this is a strange omission on Hobbes's part, and offers a proof on his behalf (1986, pp. 105–6). Vanderschraaf (2001), on the other hand, offers a reconstruction that drops Hobbes's "dubious premise" that absolute monarchy is the best form of government. Cf. also Hobbes's remark late in *Leviathan* that "which is the best [form of government of Aristotle's types] is not to be disputed where any one of them is already established. But the present ought always to be preferred, maintained, and accounted best" (L 42.82, 379), and the helpful discussion of these topics in Hoekstra, 2004.

[5] I quote the Latin passage to emphasize that Hobbes is talking about juridical *potestas*, not material *potentia*. The English passage reads: "consisteth not in the difference of power, but in the difference of convenience, or aptitude to produce

the peace and security of the people, for which end they were instituted" (L 19.4, 131). On the degree to which Hobbes separates physical and juridical power, see Foisneau, 1992 (for analysis of the terms), Matheron 1986a (for Hobbes in relation to Grotius and Spinoza) and Hoekstra 2004 (on whether Hobbes is a *de facto* or consent theorist).

6 For a discussion of this move as a reappropriation of the medieval notion of the "king's two bodies," see Pettit, 2008, pp. 70–8. See also Bredekamp, 2007, pp. 36–7.

7 Cf. again Foucault: on the standard sixteenth-century reading of Machiavelli, "the objective of the exercise of power is to reinforce, strengthen, and protect the principality, but with this last understood not to mean the objective ensemble of its subjects and the territory, but rather the prince's relation with what he owns" (1991, p. 90).

8 See also Foucault, 2003, pp. 239–54. Insofar as the sovereign executes a function, one might also speak in Foucauldian terms of "governmentality" (cf. 1991). I emphasize biopower to stress the productive aspect of the Hobbesian state, over its managerial aspects. My goal here is to establish the *prima facie* intelligibility of such a claim about biopower. In lieu of a full discussion, I will address here two preliminary objections to the propriety of the argument, in Foucauldian terms. (1) "How can Hobbes present a theory of biopower, when biopower is about the management of life, hence of nature? You just said that Hobbes banished nature from political philosophy." Answer: The formulation of the objection hides an equivocation about what "nature" means in this context. What Hobbes banishes is natural differentiation that is not subject to rational calculation. If one takes the project of biopower to be the management of "aleatory events that occur within a population that exists over a period of time" (2003, p. 246), then one way of looking at the entire Hobbesian project, of the necessity of replacing the ungovernable multitude with the regularly defined populus, and of treating the state of nature as a carefully constructed hypothesis, is precisely the removal of the aleatory. To be sure, the mechanism is different: techniques of statistics attempt to derive meaning from aleatory phenomenon at a mass level, whereas Hobbes attempts to achieve the same result by banning the aleatory from conceptual analysis, but the political objective seems to me to be the same. (2) "But Foucault is quite specific that the techniques of biopower don't emerge until the late eighteenth century. Are you challenging this thesis?" Answer: (a) A distinction should be made, I think, between actual techniques and those techniques as a desideratum. The problem of what techniques to develop, as well as what problem sets are to be articulated, is clearly not in Hobbes. On the other hand, the sharp distinction between multitude and people seems to speak to the idea of a population insofar as both indicate criteria of intelligibility and manageability. That something like techniques of biopower are desiderata seems clearly to be within the range of Hobbesian thought. (b) Foucault's chronology depends on how one articulates the relation between biopower and disciplinary power. Insofar as he associates them as flip sides of the same anti-juridical coin, and insofar as he says that techniques of discipline begin to emerge in the seventeenth century, then I think the charge that I am placing biopower 100 years too early is substantially softened. (c) Hobbes is clearly a liminal thinker in many respects,

and also anomalous within the seventeenth-century. Even if Hobbes were the theorist of biopower *par excellence*—and I do not maintain that he is—then there would still be room to talk about how the realization of the Hobbesian dream is deferred until the nineteenth century (e.g., "Hobbes influenced Bentham," a claim which I will not address here. For some skeptical thoughts, see Crimmins, 2002), and to speak of the commonplace according to which the twentieth century has seen the rise of the "new Leviathan." At the very least, I think the analysis here establishes that the spectacle of Louis XIV's absolutist court (cf. Foucault, 2003, pp. 175–7) is in no straightforward way the heir of Hobbes.

[9] Lachterman throws down the following gauntlet: after tracing a lineage from Descartes to Kant, he suggests: "The practice of philosophy in the present age continues to be governed by the etiquette of construction, now stripped of the epistemic and ontological authority Kant meant it to enjoy. In other terms, contemporary philosophy, in both its analytical and its postmodern or deconstructive versions, proceeds under Nietzschean auspices. The superficial estrangement between these two version turns out to be a family quarrel between two branches of the Nietzschean family" (1989, p. 19). The claim is perhaps extravagant, but that it is even intelligible—on the analytic side, he cites Carnap and Nelson Goodman—suggests that it bears an element of truth.

[10] Matheron calls this aspect of Hobbes an "inverse Thomism" (1986b, p. 77). See note 5, above, on the discussion of *potentia/potestas* in Hobbes.

[11] Hardt and Negri follow, as I will, with a gesture to Spinoza: "once again, Spinoza is the one who most clearly anticipates this monstrous nature of the multitude by conceiving of life as a tapestry on which the singular passions weave a common capacity of transformation, from desire to love and from the flesh to the divine body. The experience of life is for Spinoza a search for truth, perfection, and the joy of God. Spinoza shows us how today, in postmodernity, we can recognize these monstrous metamorphoses of the flesh as not only a danger but also a possibility, the possibility to create an alternative society" (194). I discuss Hardt and Negri's reception of Spinoza (at least as they present it in their earlier *Empire*) in Hull, 2005.

[12] I draw most of the references here from Schuhmann, 2004d, pp. 54–5. The following remarks are only suggestive and are intended to adumbrate one possible future course of study, from the point of view of which the present monograph should be regarded as preparatory. Among the complexities to which any thorough engagement between Hobbes and Spinoza (along the lines suggested here) must attend I will only mention two: (a) much more so than in the case of Hobbes, there is already a substantial literature (the so-called "French Materialist" reading) on Spinoza that approaches many of the issues I approach here with Hobbes. (b) There is substantial evidence of Spinoza's immersion in medieval Judeo-Islamicate thought. For some discussions of this subaltern context, see, for example, Dobbs-Weinstein, 1994 and Ravven, 2001. Hobbes may have himself encountered the Averroist Aristotle via Pomponazzi and Mothe le Vayer;

see Paganini, 2007. For representative articles from the burgeoning literature discussing the Spinoza-Hobbes connection, see Giancotti, 1997; Malcolm, 2002c; Matheron, 1986a; and Matheron, 1985. I present a very preliminary and schematic version of my own thoughts on the matter in Hull, in press.

13 I am aware that translating "salus" as "health" is an interpretive decision with which not all would agree.

Bibliography

Primary Sources

Aristotle. 1933. *Metaphysics,* trans. Hugh Tredennick. Loeb Classical Series. Cambridge, MA: Harvard UP.

—.1960. *Topics,* trans. E. S. Forster. Loeb Classical Series. Cambridge, MA: Harvard UP.

—.1980. *Physics,* trans. Hippocrates G. Apostle. Grinnell, Iowa: Peripatetic Press.

—.1981. *Posterior Analytics,* trans. Hippocrates G. Apostle. Grinnell, IA: Peripatetic Press.

—.1984. *Nichomachean Ethics,* ed. and trans. Hippocrates G. Apostle. Grinnell, Iowa: Peripatetic Press [= NE].

—.1986. *Politics,* trans. Hippocrates G. Apostle and Lloyd P. Gerson. Grinnell, Iowa: Peripatetic Press.

—.1991. *On Rhetoric,* trans. George A. Kennedy. New York: OUP.

Arnauld, Antoine and Pierre Nicole. 1996. *Logic or the Art of Thinking,* trans. Jill Vance Buroker. Cambridge: CUP.

Averrões [Ibn Rushd]. 1562. *In Porphyrium, in Aristotelis quae extant cum Opera . . . cum Averrois Cordubensis variis in eosdem commentariis.* Venice, Vol. I.

—.1963. "*The Decisive Treatise, Determining what the Connection is between Religion and Philosophy,*" trans. George F. Hourani, in *Medieval Political Philosophy: A Sourcebook,* ed. Ralph Lerner and Muhsin Mahdi. Ithaca, NY: Cornell UP, 163–87.

Avicenna [Ibn Sina]. 1508. *Opera Philosophica.* Venice. Facsimile rpt. Frankfurt: Minerva, 1961.

Bacon, Francis. 1857. *The Works of Francis Bacon,* ed. James Spedding, Robert Ellis and Douglas Heath. 13 vols. London.

—.1994. *Novum Organum,* ed. and trans. Peter Urbach and John Gibson. Chicago, IL and La Salle, IL: Open Court.

—.1996. "Of Love and Self-Love," in *Francis Bacon,* ed. Brian Vickers. Oxford: OUP.

Beza, Theodore [Théodore de Bèze]. 1970. *Du Droit des Magistrats sur leurs Sujets,* intro. and ed. Robert M. Kingdon. Geneva: Librairie Droz.

Caryl, Joseph. 1677. *An Exposition, with Practical Observations Upon the Book of Job,* 2 vols. London.

Cicero. 1952a. *De Legibus,* trans. Clinton Walker Keyes. Cambridge, MA: Harvard UP.

—.1952b. *De Re Publica,* trans. Clinton Walker Keyes. Cambridge, MA: Harvard UP.

Clarendon, Edward Hyde, Earl of. 1676. *A Brief View and Survey of the Dangerous and Pernicious Errors to Church and State, in Mr. Hobbes's Book, Entitled Leviathan.* Oxford.

Coimbra. [Collegium Conimbricensis]. 1592. *Commentarii Collegii Conimbricensis [. . .] in octo libros physicorum Aristotelis.* Coimbra. Facsimile rpt. Hildesheim, 1984.

—.1607. *Commentarii Collegii Conimbricensis e Societate Iesu: in universam dialecticam Aristotelis Stagiritae.* Cologne. Facsimile rpt. Hildesheim, 1976.

—.2001. *The Conimbricensis: Some Questions on Signs,* trans. John Doyle. Milwaukee, WI: Marquette UP.

Copernicus, Nicolas. 1995. *On the Revolutions of Heavenly Spheres,* trans. Charles Glenn Wallis. New York: Prometheus Books.

Dacres, Edward. 1673. *Nicholas Machiavel's Prince . . . with some Animadversions Noting and Taxing his Errors,* 2nd edn. London.

Descartes, René. 1964. *Oeuvres de Descartes,* ed. Charles Adam and Paul Tannery, 11 vols. Paris: J. Vrin [= AT].

—.1985. *The Philosophical Writings of Descartes,* ed. and trans. John Cottingham, Robert Stoothoff and Dugald Murdoch, 3 vols. Cambridge: CUP [= PWD].

Dowell, John. 1683. *The Leviathan Heretical: or The Charge Exhibited in Parliament against M. Hobbes, justified by the Refutation on a Book of his, Entitled The Historical Narration of Heresie and the Punishments thereof.* Oxford.

Eachard, John. 1673. *Some Opinions of Mr. Hobbs Considered in a Second Dialogue between Philautus and Timothy.* London.

—.1958. *Mr. Hobbs's State of Nature Considered, In a Dialogue between Philautus and Timothy,* ed. Peter Ure. Liverpool: Liverpool UP.

Eustachius à Sancto Paulo. 1620. *Summa Philosophica Quadripartita, De rebus Dialecticis, Ethicis, Physicis, & Metaphysicis.* Paris.

Filmer, Sir Robert. 1991. "*Observations Concerning the Original of Government, upon Mr Hobbes's 'Leviathan', Mr Milton against Salmasius, and H. Grotius De Jure Belli,*" in *Patriarcha and Other Writings,* ed. Johann P. Sommerville. Cambridge: CUP, 184–234.

Gentillet, Innocent. 1968. *Anti-Machiavel,* ed. Edward Rathé. Geneva: Librairie Droz.

Gersonides, Levi ben. 1562. *In Porphyrium, in Aristotelis quae extant cum Opera . . . cum Averrois Cordubensis variis in eosdem commentariis.* Venice, Vol. I.

Goulston, Theodore. 1619. *Aristotelis de Rhetorica seu arte Dicendi Libri Tres.* London.

Grotius, Hugo. 1609. Mare Liberum. Amsterdam.

—.1962. *The Law of War and Peace. De jure belli ac pacis libri tres,* trans. Francis W. Kelsey. Indianapolis, IN: Bobbs-Merrill.

Harrington, James. 1992. *The Commonwealth of Oceana and a System of Politics,* ed. J. G. A. Pocock. Cambridge: CUP.

Hobbes, Thomas. 1839. *Opera Philosophica quae latine scripsit omnia,* ed. William Molesworth, 5 vols. London [= OL].

—.1839–45. *The English Works of Thomas Hobbes,* ed. Sir William Molesworth, 11 vols. London [= EW].

—.1971. *A Dialogue between a Philosopher and a Student of the Common Laws of England,* ed. Joseph Cropsey. Chicago, IL: U. Chicago Press.

—.1973. *De Motu*, as *Critique du De Mundo de Thomas White*, ed. Jean Jacquot and Harold Whitmore Jones. Paris: J. Vrin.
—.1983. *De Cive, The Latin Version*, ed. Howard Warrender. Oxford: Clarendon Press [= DC].
—.1990. *Behemoth, or the Long Parliament*, ed. Ferdinand Tönnies. Chicago, IL: U. Chicago Press.
—.1991a. *De Homine*, in *Man and Citizen*, ed. Bernard Gert., IN: Hackett [= DH].
—.1991b. *Leviathan*, ed. Richard Tuck. Cambridge: CUP [= L].
—.1994a. *The Correspondence of Thomas Hobbes*, ed. Noel Malcolm, 2 vols. Oxford: Clarendon Press [= HC].
—.1994b. *Elements of Law, Natural and Politic*, rpt. as *Human Nature and De Corpore Politico*, ed. J. C. A. Gaskin. Oxford: OUP [= EL].
—.1994c. *Leviathan, with Selected Variants from the Latin Edition of 1668*, ed. Edwin Curley. Indianapolis, IN: Hackett [= L].
—.1998. *On the Citizen [De Cive]*, ed. and trans. Richard Tuck and Michael Silverstone. Cambridge: CUP [= DC].
—.1999a. *De Corpore*, ed. Karl Schuhmann. Paris: J. Vrin [= DCo].
—.1999b. "Of Liberty and Necessity," in *Hobbes and Bramhall on Liberty and Necessity*, ed. Vere Chappell. Cambridge: CUP, 15–42.
Homer. 1924. *The Iliad*, trans. A. T. Murray, revised William F. Wyatt. Cambridge, MA., Harvard UP.
Kant, Immanuel. 1991. "Idea for a Universal History with a Cosmopolitan Purpose," in *Kant: Political Writings*, trans. H. B. Nisbet, ed. Hans Reiss, 2nd edn. Cambridge: CUP, 41–53.
Kepler, Johannes. 1856–71. *Opera omnia*, ed. Christian Frisch, 8 vols. Frankfurt.
Lactantius. 1871. "*The Divine Institutes*," in *The Ante-Nicene Fathers* Vol. 7, ed. Alexander Roberts and James Donaldson. Grand Rapids, MI: Wm. B. Eerdmans.
Leibniz, G. W. 1686. "Discourse on Metaphysics," in *Philosophical Texts*, trans. R. S. Woolhouse and Richard Francks. Oxford: OUP, 53–94.
—.1875–1890. *Philosophische Schriften*, ed. Gerhardt. Berlin.
Locke, John. 1974. *An Essay Concerning Human Understanding*, ed. Peter H. Nidditch. Oxford, Clarendon Press.
—.1988. *Two Treatises of Government*, ed. Peter Laslett. Cambridge: CUP [= 2T].
—.1997. *Political Essays*, ed. Mark Goldie. Cambridge: CUP.
Machiavelli, Niccolò. 1965. *The Chief Works and Others*, trans. Allan Gilbert, 3 vols. Durham, NC: Duke UP [= W].
—.1988. *The Prince*, trans. Russell Price and ed. Quentin Skinner. Cambridge: CUP [= P].
—.1996. *Discourses on Livy*, trans. Harvey C. Mansfield and Nathan Tarcov. Chicago, IL: U. Chicago Press.
Maimonides, Moses. 1963. *The Guide of the Perplexed*, trans. Shlomo Pines, 2 vols. Chicago, IL: U. Chicago Press.
Marx, Karl. 1927. *Die heilige Familie, oder Kritik der kritischen Kritik.* MEGA[1] I/3, Berlin and Moscow, 175–390.
Philolaus. 1993. *Philolaus of Croton: Pythagorean and Presocratic*, ed. Carl A. Huffman. Cambridge: CUP.

Philostratus. 1960. *Life of Appolonius of Tyana,* trans. F. C. Conybeare. Cambridge, MA: Harvard UP.

Plato. 1961. *The Collected Dialogues,* ed. Edith Hamilton and Huntington Cairns. Princeton, NJ: Princeton UP.

Porphyry. 2003. *Introduction* [*Isagoge*], trans. Jonathan Barnes. Oxford: OUP.

Quintilian. 2001. *Institutio Oratoria,* trans. Donald A. Russell. Cambridge, MA: Harvard UP.

Rousseau, Jean Jacques. 1973. *The Social Contract and Discourses,* trans. G. D. H. Cole. London: Everyman Library.

Spinoza, Baruch. 1925. *Ethica ordine geometrico demonstrata,* in *Opera II,* ed. Carl Gebhardt. Heidelberg: Carl Winter, 45–308.

St. Thomas Aquinas. 1948. *Summa Theologiae,* trans. Fathers of the English Dominican Province, 3 vols. New York: Benzinger Bros.

Suárez, Francisco. 1947. *On the Various Kinds of Distinctions,* trans. Cyril Vollert. Milwaukee, WI: Marquette UP.

Wallis, John. 1657. *Mathesis Universalis,* in *Opera Mathematica.* Oxford.

—.1662. *Hobbius heauton-timorumenos.* London.

—. 1687. *Institutio Logicae.* Oxford.

Whitehall, John. 1680. *Behemoth Arraign'd: or, a Vindication of Property Against a Fanatical Pamphlet Stiled Behemoth: or, the History of the Civil Wars of England, from 1640 to 1660.* London.

William of Ockham. 1967–86. *Opera Theologica,* 10 vols. New York: St. Bonaventure.

—.1974a. *Summa Logicae,* in *Opera Philosophica,* Vol. I. New York: St. Bonaventure [= SL].

—.1974b. *Ockham's Theory of Terms: Part I of the Summa Logicae,* trans. Michael J. Loux. Notre Dame, IN: U. Notre Dame Press [= SL].

Zabarella, Jacopo. 1597. *Opera Logica.* Köln, ed. Wilhelm Risse. Facsimile rpt. Hildesheim, 1966.

Secondary Sources

Agamben, Giorgio. 1998. *Homo Sacer: Sovereign Power and Bare Life,* trans. Daniel Heller-Roazen. Stanford, CA: Stanford UP.

Alféri, Pierre. 1989. *Guillaume d'Ockham: Le Singulier.* Paris: Les Éditions de Minuit.

Ariew, Roger. 1989. *Descartes and the Last Scholastics.* Ithaca, NY: Cornell UP, 1999.

Ashcraft, Richard. 1971. "Hobbes's Natural Man: A Study in Ideology Formation," *Journal of Politics* 33, 1076–1117.

—. 1986. *Revolutionary Politics and Locke's Two Treatises of Government.* Princeton, NJ: Princeton UP.

Baker, C. Edwin. 1975. "The Ideology of the Economic Analysis of Law," *Philosophy and Public Affairs* 5, 3–48.

Balibar, Étienne. 1996. "Sujet, individu, citoyen. Qu'est-ce que 'l'homme' au XVIIe siècle?" In *L'individu dans la théorie politique et dans la pratique,* ed. Janet Coleman. Paris: PUF, 249–77.

Ball, Terence. 1987. "In the Shadow of Babel: the 'Scientific' Reconstruction of Political Discourse," in *Political Discourse*, ed. Bhikhu Parekh. New Delhi: Sage, 23–46.

—.1995. "Hobbes's Linguistic Turn," in *Reappraising Political Theory: Revisionist Studies in the History of Political Thought*. Oxford: Clarendon Press, 83–106.

Barrachi, Claudia. 2003. "The Nature of Reason and the Sublimity of First Philosophy: Towards a Reconfiguration of Aristotelian Interpretation," *Epoché* 7, 223–50.

Baumgold, Deborah. 2004. "The Composition of Hobbes's *Elements of Law*," *History of Political Thought* 25, 16–43.

Bell, David R. 1969. "What Hobbes Does with Words," *Philosophical Quarterly* 19, 155–8.

Bertman, Martin A. 1990. "Hobbes and Xenophon's *Tyrannicus*," in *Thomas Hobbes: Le Ragioni del Moderno tra Teologia e Politica*, ed. Gianfranco Borrelli. Naples: Morano Editore, 129–45.

Bianchi, Emanuela. 2006. "Material Vicissitudes and Technical Wonders: The Ambiguous Figure of *Automaton* in Aristotle's Metaphysics of Sexual Difference," *Epoché* 11, 109–40.

Biletzki, Anat. 1997. *Talking Wolves: Thomas Hobbes on the Language of Politics and the Politics of Language*. Dordrecht: Kluwer.

Bird, Alexander. 1996. "Squaring the Circle: Hobbes on Philosophy and Geometry," *Journal of the History of Ideas* 57, 217–31.

Blair, Bruce G. 1993. *The Logic of Accidental Nuclear War*. Washington: Brookings Institute.

Blits, Jan H. 1989. "Hobbesian Fear," *Political Theory* 17, 417–31.

Boler, John F. 1982. "Intuitive and Abstractive Cognition," in *The Cambridge History of Later Medieval Philosophy*, ed. Norman Kretzmann, Anthony Kenny and Jan Pinborg. Cambridge: CUP, 460–78.

Borrelli, Gianfranco. 1997. "Obligation juridique et obéissance politique: les temps de la discipline moderne pour Jean Bodin, Giovanni Botero et Thomas Hobbes," in *Politique, Droit et Théologie chez Bodin, Grotius et Hobbes*, ed. Yves Charles Zarka. Paris: Éditions Kimé, 11–25.

Boyle, James. 1986. "Thomas Hobbes and the Invented Tradition of Positivism: Reflections on Language, Power, and Essentialism," *University of Pennsylvania Law Review* 135, 383–426.

Bray, Michael. 2006. "The Hedges that Are Set: Hobbes and the Future of Politics," *Epoché* 11, 173–200.

—.2007. "Macpherson Restored? Hobbes and the Question of Social Origins," *History of Political Thought* 28, 56–90.

Bredekamp, Horst. 2007. "Thomas Hobbes's Visual Strategies," in *The Cambridge Companion to Hobbes's Leviathan*, ed. Patricia Springborg. Cambridge: CUP, 29–60.

Breiner, Peter. 2008. "Machiavelli's 'New Prince' and the Primordial Moment of Acquisition," *Political Theory* 36, 66–92.

Brown, Clifford W., Jr. 1987. "Thucydides, Hobbes, and the Derivation of Anarchy," *History of Political Thought* 8, 33–62.

—.1989. "Thucydides, Hobbes, and the Linear Causal Perspective," *History of Political Thought* 10, 215–56.

Burke, Peter. 2003. "Images as Evidence in Seventeenth-Century Europe," *Journal of the History of Ideas* 64, 273–96.

Calabresi, Guido and A. Douglas Melamed. 1972. "Property Rules, Liability Rules, and Inalienability: One View of the Cathedral," *Harvard Law Review* 85, 1089–1128.

Callaghan, G. K. 2001. "Nominalism, Abstraction, and Generality in Hobbes," *History of Philosophy Quarterly* 18, 37–55.

Carter, Jennifer. 1979. "Law, Courts and Constitution," in *The Restored Monarchy 1660–1688*, ed. J. R. Jones. New Jersey: Rowan and Littlefield, 71–93.

Chanteur, Janine. 1969. "*Note sur les Notions de 'Peuple' et de 'Multitude' chez Hobbes*," in *Hobbes-Forschungen*, ed. Reinhart Koselleck and Roman Schnur. Berlin: Duncker & Humboldt, 223–35.

Coase, R. H. 1960. "The Problem of Social Cost," *Journal of Law and Economics* 3, 1–44.

Colish, Marcia L. 1999. "Republicanism, Religion, and Machiavelli's Savonarolan Moment," *Journal of the History of Ideas* 60, 597–616.

Copp, David. 1980. "Hobbes on Artificial Persons and Collective Actions," *Philosophical Review* 89, 579–606.

Cover, Robert M. 1986. "Violence and the Word," *Yale Law Journal* 96, 1601–29.

Crimmins, James E. 2002. "Bentham and Hobbes: An Issue of Influence," *Journal of the History of Ideas* 63, 677–96.

Curley, Edwin. 1989. "Reflections on Hobbes: Recent Work on his Moral and Political Philosophy," *Journal of Philosophical Research* 15, 169–250.

—.1996. "Calvin and Hobbes, or, Hobbes as an Orthodox Christian," *Journal of the History of Philosophy* 34, 257–71.

Curran, Elanor. 2002. "A Very Peculiar Royalist: Hobbes in the Context of His Political Contemporaries," *British Journal for the History of Philosophy* 10, 167–208.

D'Amico, Jack. 1987. "Machiavelli's Borgia: Founder and Failure," *Rivista di Studi Italiani* 5–6: 2–1, 18–30.

Davidson, J. S. 1994. "The Rights of Indigenous People in Early International Law," *Canterbury Law Review* 5, 391–423.

Dear, Peter. 1995. *Discipline and Experience: The Mathematical Way in the Scientific Revolution.* Chicago, IL: U. Chicago Press.

—.2001. *Revolutionizing the Sciences: European Knowledge and its Ambitions, 1500–1700.* Princeton, NJ: Princeton UP.

Deleuze, Gilles. 1994. *Difference and Repetition*, trans. Paul Patton. New York: Columbia UP.

Des Chene, Dennis. 1996. *Physiologia: Natural Philosophy in Late Aristotelian and Cartesian Thought.* Ithaca, NY: Cornell UP.

Dobbs-Weinstein, Idit. 1994. "Maimonidean Aspects in Spinoza's Thought," *Graduate Faculty Philosophy Journal* 17:1–2, 153–74.

—.1995. *Maimonides and St. Thomas on the Limits of Reason.* Albany, NY: SUNY Press.

Farneti, Roberto. 2001. "The 'Mythical Foundation' of the State: Leviathan in Emblematic Context," *Pacific Philosophical Quarterly* 82, 362–82.

Femia, Joseph V. 1988. "An Historicist Critique of 'Revisionist' Methods for Studying the History of Ideas," in *Meaning and Context: Quentin Skinner and His Critics*, ed. James Tully. Princeton, NJ: Princeton UP, 156–75.

Foisneau, Luc. 1992. "*Le vocabulaire du pouvoir: Potentia/Potestas,* Power," in *Hobbes et son Vocabulaire,* ed. Yves Charles Zarka. Paris: J. Vrin, 83–102.

—.1997. "Hobbes et la théorie machiavélienne de la virtù," *Archives de Philosophie* 60, 371–91.

Forster, Greg. 2003. "Divine Law and Human Law in Hobbes's *Leviathan,*" *History of Political Thought* 24, 189–217.

Foucault, Michel. 1970. *The Order of Things: An Archaeology of the Human Sciences.* New York: Random House.

—.1978. *The History of Sexuality Vol I: An Introduction,* trans. Robert Hurley. New York: Vintage Books.

—.1991. "Governmentality," in *The Foucault Effect: Studies in Governmentality,* ed. Graham Burchell, Colin Gordon and Peter Miller. Chicago, IL: U. Chicago Press, 87–104.

—.2003. "*Society Must be Defended:" Lectures at the Collège de France, 1975–76,* trans. David Macey. New York: Picador.

Gauthier, David. 1977. "The Social Contract as Ideology," *Philosophy and Public Affairs* 6, 130–64.

—.1987. "Taming Leviathan," *Philosophy and Public Affairs* 16, 280–98.

—.1988. "Hobbes's Social Contract," in *Perspectives on Thomas Hobbes,* ed. G. A. J. Rogers and Alan Ryan. Oxford: Clarendon Press, 125–52.

—.1997. "Hobbes on Demonstration and Construction," *Journal of the History of Philosophy* 35, 509–21.

Giancotti, Emilia. 1997. "The Birth of Modern Materialism in Hobbes and Spinoza," in *The New Spinoza,* ed. Warren Montag and Ted Stolze. Minneapolis, MN: U. Minnesota Press, 49–63.

Goodhart, Michael. 2000. "Theory in Practice: Quentin Skinner's Hobbes, Reconsidered," *Review of Politics* 62, 531–61.

Goyard-Fabre, Simone. 1982. "Right and Anthropology in Hobbes's Philosophy," in *Thomas Hobbes: His View of Man,* ed. J. G. van der Bend. Amsterdam: Editions Rodopi, 17–30.

—.1995. "De la philosophie première à la philosophie politique dans la pensée de Thomas Hobbes," in *Politica e Diritto in Hobbes,* ed. Giuseppe Sorgi. Milan: Dott. A. Giuffrè Editore, 75–95.

Greene, Thomas M. 1986. "The End of Discourse in Machiavelli's *Prince,*" in *Literary Theory/Renaissance Texts,* ed. Patricia Parker and David Quint. Baltimore, MD and London: Johns Hopkins UP, 63–77.

Hacking, Ian. 1975. "Thomas Hobbes's Mental Discourse," in *Why Does Language Matter to Philosophy?* Cambridge: CUP, 15–25.

Hadfield, Andrew. 2001. "The Name 'Philautus' in Barnaby Rich and John Lyly," *Notes and Queries,* September, 313–4.

Hampton, Jean. 1985. "Hobbes's State of War," *Topoi* 4, 47–60.

—. 1986. *Hobbes and the Social Contract Tradition.* Cambridge: CUP.

—. 1991. "Cooperating and Contracting: A Reply to I. Haji," *Philosophy and Phenomenological Research* 51, 603–9.

Hanson, Donald W. 1991. "Reconsidering Hobbes' Conventionalism," *Review of Politics* 53, 627–51.

—.1993. "Science, Prudence and Folly in Hobbes's Political Theory," *Political Theory* 21, 643–64.

Haraway, Donna J. 1991. "A Cyborg Manifesto: Science, Technology and Socialist-Feminism in the Late Twentieth Century," in *Simians, Cyborgs and Women: The Reinvention of Nature*. New York: Routledge, 149–81.

Hardt, Michael and Antonio Negri. 2000. *Empire*. Cambridge, MA: Harvard UP.

—.2004. *Multitude: War and Democracy in the Age of Empire*. New York: Penguin.

Hariman, Robert. 1989. "Composing Modernity in Machiavelli's *Prince*," *Journal of the History of Ideas* 50, 3–29.

Hill, Christopher. 1997. *The Intellectual Origins of the English Revolution Revisited*. Oxford: Clarendon Press.

Hoekstra, Kinch. 1997. "Hobbes and the Foole," *Political Theory* 25, 620–54.

—.1999. "Nothing to Declare? Hobbes and the Advocate of Injustice," *Political Theory* 27, 230–5.

—.2001. "Tyrannus Rex vs. Leviathan," *Pacific Philosophical Quarterly* 82, 420–46.

—.2004. "The *de facto* turn in Hobbes's Political Philosophy," in *Leviathan after 350 Years*, ed. Tom Sorrell and Luc Foisneau. Oxford: Clarendon Press, 33–74.

—.2007. "Hobbes on the Natural Condition of Mankind," in *The Cambridge Companion to Hobbes's Leviathan*, ed. Patricia Springborg. Cambridge: CUP, 109–27.

Huffman, Carl A. 1993. "Introduction" in *Philolaus of Croton: Pythagorean and Presocratic*, ed. Carl A. Huffman. Cambridge: CUP.

—.2003. *Philolaus of Croton: Pythagorean and Presocratic*, ed. Carl A. Huffman. Cambridge: CUP.

Hull, Gordon. 2002. "'Against this *Empusa*': Hobbes's *Leviathan* and the Book of Job," *British Journal for the History of Philosophy* 10, 3–29.

—.2005. "*Capital sive Natura*: Spinoza and the Immanence of *Empire*," *International Studies in Philosophy* 37, 15–35.

—.in press. "Loving Well: Affective Economics in Hobbes, Locke, and Spinoza," *NASS Monograph*.

Hungerland, Isabel C. and George R. Vick. 1973. "Hobbes's Theory of Signification," *Journal of the History of Philosophy* 11, 459–82.

Jardine, Nicholas. 1988. "Epistemology of the Sciences," in *The Cambridge History of Renaissance Philosophy*, ed. Charles B. Schmitt, Quentin Skinner and Eckhard Kessler. Cambridge: CUP, 685–711.

Jesseph, Douglas M. 1999a. "The Decline and Fall of Hobbesian Geometry," *Studies in the History and Philosophy of Science* 30, 425–53.

—.1999b. *Squaring the Circle: The War between Hobbes and Wallis*. Chicago, IL: U. Chicago Press.

—.2002. "Hobbes's Atheism," *Midwest Studies in Philosophy* 26, 140–66.

—.2006. "Hobbesian Mechanics," *Oxford Studies in Early Modern Philosophy* 3, 119–52.

—.unpubl. ms. "Geometry, Theology, and Politics: Context and Consequences of the Hobbes-Wallis Dispute."

Johnson, Paul J. 1982. "Hobbes and the Wolf-Man," in *Thomas Hobbes: His View of Man*, ed. J. G. van der Bend. Amsterdam: Editions Rodopi, 31–44.

Johnston, David. 1986. *The Rhetoric of Leviathan: Thomas Hobbes and the Politics of Cultural Transformation*. Princeton, NJ: Princeton UP.

Jones, W. J. 1971. *Politics and the Bench: The Judges and the Origins of the English Civil War*. London: George Allen and Unwin and New York: Barnes and Noble.

Kahn, Victoria. 2001. "Hobbes, Romance, and the Contract of Mimesis," *Political Theory* 29, 4–29.

Kaplan, Robert D. 1994. "The Coming Anarchy," *The Atlantic*, February, 44–76.

Kavka, Gregory S. 1983. "Hobbes's War of All against All," *Ethics* 93, 291–310.

Keane, John. 1988. "More Theses on the Philosophy of History," in *Meaning and Context: Quentin Skinner and His Critics*, ed. James Tully. Princeton, NJ: Princeton UP, 204–17.

Kilgour, D. Marc and Frank C. Zagare. 1991. "Credibility, Uncertainty and Deterrence," *American Journal of Political Science* 35, 305–34.

Klein, Jacob. 1968. *Greek Mathematical Thought and the Origin of Algebra*, trans. Eva Brann. New York: Dover.

Klosko, George and Daryl Rice. 1985. "Thucydides and Hobbes's State of Nature," *History of Political Thought* 6, 405–9.

Kottman, Karl A. 1975. "Fray Luis de León and the Universality of Hebrew: An Aspect of 16th and 17th Century Language Theory," *Journal of the History of Philosophy* 13, 297–310.

Lachterman, David R. 1989. *The Ethics of Geometry: A Genealogy of Modernity*. New York: Routledge.

Largeault, Jean. 1971. *Enquête sur le Nominalisme*. Paris: Béatrice-Nauwelaerts.

Lazzeri, Christian. 1990. "Les racines de la volonté de puissance: le 'passage' de Machiavel à Hobbes," in *Thomas Hobbes: Philosophie première, théorie de la science et politique*, ed. Yves Charles Zarka and Jean Bernhardt. Paris: PUF, 235–46.

Lee, Richard A. 2000. "The Play of Force versus the Reduction of Force: Hobbes and Roger Bacon on Perception," *Hobbes Studies* 13, 34–45.

—.2002. *Science, The Singular, and The Question of Theology*. New York: Palgrave.

—.2004. *The Force of Reason and the Logic of Force*. New York: Palgrave.

Leijenhorst, Cees. 2001. *The Mechanisation of Aristotelianism: The Late Aristotelian Setting of Hobbes's Natural Philosophy*. Leiden: Brill.

—.2007. "Sense and Nonsense about Sense: Hobbes and the Aristotelians on Sense Perception and Imagination," in *The Cambridge Companion to Hobbes's Leviathan*, ed. Patricia Springborg. Cambridge: CUP.

Levy, M. 1995. "Is the Environment a National Security Issue?" *International Security* 20:2, 35–62.

Lloyd, S. A. 2001. "Hobbes's Self-Effacing Natural Law Theory," *Pacific Philosophical Quarterly* 82, 285–308.

Ludwig, Bernd. 1995. "Scientia civilis more geometrico—Die philosophische Methode als architektonisches Prinzip in Hobbes' Leviathan," *Hobbes Studies* 8, 46–87.

MacDonald, Paul S. 2002. "Descartes: The Lost Episodes," *JHP* 40, 437–60.

Malcolm, Noel. 2002a. "Charles Cotton, Translator of Hobbes's *De Cive*," in *Aspects of Hobbes*. Oxford: Clarendon Press, 234–58.

—.2002b. "Hobbes and the European Republic of Letters," in *Aspects of Hobbes*. Oxford: Clarendon Press, 457–545.

—.2002c. "Hobbes and Spinoza," in *Aspects of Hobbes*. Oxford: Clarendon Press, 27–52.

—.2002d. "Hobbes, Ezra, and the Bible: The History of a Subversive Idea," in *Aspects of Hobbes*. Oxford: Clarendon Press, 383–431.

—.2002e. "Hobbes, Sandys, and the Virginia Company," in *Aspects of Hobbes*. Oxford: Clarendon Press, 53–79.

—.2002f. "Hobbes's Theory of International Relations," in *Aspects of Hobbes*. Oxford: Clarendon Press, 432–56.

—.2002g. "The Title Page of *Leviathan*, Seen in a Curious Perspective," in *Aspects of Hobbes*. Oxford: Clarendon Press, 200–33.

Malet, Antoni. 2001. "The Power of Images: Mathematics and Metaphysics in Hobbes's Optics," *Studies in the History and Philosophy of Science* 32, 303–33.

Mancosu, Paolo. 1992. "Aristotelian Logic and Euclidean Mathematics: Seventeenth-Century Developments of the *Quaestio do certitudine mathematicarum*," *Studies in the History and Philosophy of Science* 23, 241–65.

Martinich, A. P. 2001. "Interpretation and Hobbes's Political Philosophy," *Pacific Philosophical Quarterly* 82, 309–31.

Matheron, Alexandre. 1985. "Le 'droit le plus fort': Hobbes contre Spinoza," *Revue philosophique de la France et de l'étranger* 2, 149–76.

—.1986a. "Spinoza et la problématique juridique de Grotius," in *Anthropologie et Politique au XVIIe Siecle*. Paris: J. Vrin, 81–101.

—.1986b. "Spinoza et la decomposition de la Politique thomiste: Machiavélisme et Utopie," in *Anthropologie et Politique au XVIIe Siecle*. Paris: J. Vrin, 49–79.

Maynes, Charles William. Fall, 1995. "The New Pessimism," *Foreign Policy* 100, 33–49.

McPherson, C. B. 1962. *The Political Theory of Possessive Individualism*. Oxford: Clarendon Press.

Menn, Stephen. 1997. "Suárez, Nominalism, and Modes," in *Hispanic Philosophy in the Age of Discovery*, ed. Kevin White. Washington, DC: Catholic University of America Press, 226–56.

Milton, J. R. 1987. "Induction before Hume," *British Journal for the Philosophy of Science* 38, 49–74.

Miner, Robert C. 2001. "Is Hobbes a Theorist of the Virtues?" *International Philosophical Quarterly* 41, 269–84.

Mintz, Samuel L. 1962. *The Hunting of Leviathan: Seventeenth-Century Reactions to the Materialism and Moral Philosophy of Thomas Hobbes*. Cambridge: CUP.

—.1989. "Leviathan as Metaphor," *Hobbes Studies* 2, 3–9.

Missner, Marshall. 1983. "Skepticism and Hobbes's Political Philosophy," *Journal of the History of Ideas* 44, 407–27.

Nederman, Cary J. 1994. "The Puzzle of the Political Animal: Nature and Artifice in Aristotle's Political Theory," *Review of Politics* 56, 282–304.

—.1996. "The Meaning of 'Aristotelianism' in Medieval Moral and Political Thought,' *Journal of the History of Ideas* 57, 563–85.

—.1999. "Amazing Grace: Fortune, God, and Free Will in Machiavelli's Thought," *Journal of the History of Ideas* 60, 617–38.

Newman, William. 1989. "Technology and the Alchemical Debate in the Late Middle Ages," *Isis* 80, 423–45.

Normore, Calvin G. 1997. "Material Supposition and the Mental Language of Ockham's *Summa Logicae*," *Topoi* 16, 27–33.

Pacchi, Arrigo. 1988. "Hobbes and the Problem of God," in *Perspectives on Thomas Hobbes*, ed. G. A. J. Rogers and Alan Ryan. Oxford: Clarendon Press, 171–88.

Paganini, Gianni. 2007. "Hobbes's Critique of the Doctrine of Essences and its Sources," in *The Cambridge Companion to Hobbes's Leviathan*, ed. Patricia Springborg. Cambridge: CUP, 337–57.

Parry, Geraint. 1967. "Performative Utterances and Obligation in Hobbes," *Philosophical Quarterly* 17, 246–52.

Pasquino, Pasquale. 1994. "Thomas Hobbes: La condition naturelle de l'humanité," *Revue française de science politique* 44, 294–307.

—.2001. "Hobbes, Religion, and Rational Choice: Hobbes's Two Leviathans and the Fool," *Pacific Philosophical Quarterly* 82, 406–19.

Patapan, Haig. 2000. "'Lord Over the Children of Pride': The Vaine-Glorious Rhetoric of Hobbes's Leviathan," *Philosophy and Rhetoric* 33, 74–93.

Pateman, Carole. 1988. *The Sexual Contract.* Stanford, CA: Stanford UP.

Pérez-Ramos, Antonio. 1988. *Francis Bacon's Idea of Science and the Maker's Knowledge Tradition.* Oxford: Clarendon Press.

Pesic, Peter. 2000. "Kepler's Critique of Algebra," *The Mathematical Intelligencer* 22:4, 54–9.

Pettit, Philip. 2008. *Made with Words: Hobbes on Language, Mind and Politics.* Princeton, NJ: Princeton UP.

Pincombe, Mike. 2002. "'Philautus' in Rich and Lyly: A Suspicion Confirmed," *Notes and Queries,* June, 253–4.

Pocock, J. G. A. 1975. *The Machiavellian Moment: Florentine Political Thought and the Atlantic Republican Tradition.* Princeton, NJ: Princeton UP.

Polin, Raymond. 1995. "Les 'lois de nature' à la mode Hobbes," in *Politica e Diritto in Hobbes,* ed. Giuseppe Sorgi. Milan: Dott. A. Giuffrè Editore, 53–62.

Popkin, Richard H. 1964. *The History of Skepticism from Erasmus to Descartes,* revised edn. New York: Harper Torchbooks.

Prins, Jan. 1990. "Hobbes and the School of Padua: Two Incompatible Approaches to Science," *Archiv für Geschichte der Philosophie* 22, 26–46.

Probst, Siegmund. 1993. "Infinity and Creation: the Origin of the Controversy between Thomas Hobbes and the Savilian Professors Seth Ward and John Wallis," *British Journal for the History of Science* 26, 271–9.

Pycior, Helena M. 1997. *Symbols, Impossible Numbers, and Geometric Entanglements: British Algebra through the Commentaries on Newton's Universal Arithmetick.* Cambridge: CUP.

Raab, Felix. 1964. *The English Face of Machiavelli: A Changing Interpretation, 1500–1700.* London: Routledge and Kegan Paul.

Rahe, Paul A. 2007. "In the Shadow of Lucretius: The Epicurean Foundations of Machiavelli's Political Thought," *History of Political Thought* 28, 30–55.

Rancière, Jacques. 1999. *Disagreement: Politics and Philosophy,* trans. Julie Rose. Minneapolis, MN: U. Minnesota Press.

Ravven, Heidi M. 2001. "Some Thoughts on what Spinoza Learned from Maimonides on the Prophetic Imagination Part Two: Spinoza's Maimonideanism," *Journal of the History of Philosophy* 39, 385–406.

Reik, Miriam M. 1977. *The Golden Lands of Thomas Hobbes.* Detroit: Wayne State UP.

Reiss, Timothy J. 1982. *The Discourse of Modernism.* Ithaca, NY and London: Cornell UP.

—.1991. "Descartes, the Palatinate, and the Thirty Years War: Political Theory and Political Practice," in *Baroque Topographies,* ed. Timothy Hampton, *Yale French Studies* 80. New Haven, CT: Yale UP, 108–45.

—.1992. *The Meaning of Literature.* Ithaca, NY and London: Cornell UP.

—.1997. *Knowledge, Discovery and Imagination in Early Modern Europe.* Cambridge: CUP.

—.2000. "Neo-Aristotle and Method: Between Zabarella and Descartes," in *Descartes' Natural Philosophy,* ed. Stephen Gaukroger, John Schuster and John Sutton. New York: Routledge, 195–227.

—.2004. "Calculating Humans: Mathematics, War, and the Colonial Calculus," in *Arts of Calculation: Quantifying Thought in Early Modern Europe,* ed. David Glimp and Michelle R. Warren. New York: Palgrave/Macmillan, 137–64.

Richir, Marc. 1997. "Doute hyperbolique et 'Machiavélism:' L'institution du sujet moderne chez Descartes," *Archives de Philosophie* 60, 109–22.

Rogers, G. A. J. 1988. "Hobbes's Hidden Influence," in *Perspectives on Thomas Hobbes,* ed. G. A. J. Rogers and Alan Ryan. Oxford: Clarendon Press, 189–205.

—.1990. "Religion and the Explanation of Action in the Thought of Thomas Hobbes," in *Thomas Hobbes: Le Ragioni del Moderno tra Teologia e Politica,* ed. Gianfranco Borrelli. Naples: Morano Editore, 35–50.

Rose, Paul Lawrence and Stillman Drake. 1971. "The Pseudo-Aristotelian Questions of Mechanics in Renaissnace Culture," *Studies in the Renaissance* 18, 65–104.

Ross, George MacDonald. 1988. "Hobbes and Descartes on the Relation between Language and Consciousness," *Synthese* 75, 217–29.

Rossi, Paolo. 2000. *Logic and the Art of Memory: The Quest for a Universal Language,* trans. Stephen Clucas. Chicago, IL: U. Chicago Press.

Röttgers, Kurt. 1975. *Kritik und Praxis: Zur Geschichte des Kritikbegriffs von Kant bis Marx.* Berlin: Walter de Gruyter.

Roux, Louis. 1990. "Modelés théologiques et modelés scientifiques dans la pensée de Thomas Hobbes," in *Thomas Hobbes: Le Ragioni del Moderno tra Teologia e Politica,* ed. Gianfranco Borrelli. Naples: Morano Editore, 83–94.

Runciman, David. 1997. *Pluralism and the Personality of the State.* Cambridge: CUP.

—.2000. "What Kind of Person is Hobbes's State? A Reply to Skinner," *Journal of Political Philosophy* 8, 268–78.

Sacksteder, William. 1980. "Hobbes: The Art of the Geometricians," *Journal of the History of Philosophy* 18, 131–46.

—.1981a. "Hobbes: Geometrical Objects," *Philosophy of Science* 48, 573–90.

—.1981b. "Some Ways of Doing Language Philosophy: Nominalism, Hobbes, and the Linguistic Turn," *Review of Metaphysics* 34, 459–85.

—.1982. "Hobbes: Man the Maker," in *Thomas Hobbes: His View of Man,* ed. J. G. van der Bend. Amsterdam: Editions Rodopi, 77–88.

Sallis, John. 1996. "The Politics of the *Chora,*" in *The Ancients and the Moderns,* ed. Reginald Lilly. Bloomington, IN and Indianapolis, IN: Indiana UP, 59–71.

Schlag, Pierre. 1989. "The Problem of Transaction Costs," *Southern California Law Review* 62, 1661–700.

Schlatter, Richard. 1945. "Thomas Hobbes and Thucydides," *Journal of the History of Ideas* 6, 350–62.

Schmaltz, Tad M. 2002. *Radical Cartesianism: The French Reception of Descartes.* Cambridge: CUP.

Schuhmann, Karl. 1985. "Geometrie und Philosophie bei Thomas Hobbes," *Philosophisches Jahrbuch* 92, 161–77.

—.1990. "Hobbes and Aristotle's *Politics*," in *Thomas Hobbes: Le Ragioni del Moderno tra Teologia e Politica*, ed. Gianfranco Borrelli. Naples: Morano Editore, 97–127.

—.1998. "Skinner's Hobbes," *British Journal for the History of Philosophy* 6, 115–25.

—.2004a. "Hobbes and the Political Thought of Plato and Aristotle," in *Selected Papers on Renaissance Philosophy and on Thomas Hobbes*, ed. Piet Steenbakkers and Cees Leijenhorst. Dordrecht: Kluwer, 191–218.

—.2004b. "Hobbes und Gassendi," in *Selected Papers on Renaissance Philosophy and on Thomas Hobbes*, ed. Piet Steenbakkers and Cees Leijenhorst. Dordrecht: Kluwer, 219–25.

—.2004c. "La notion de loi chez Hobbes," in *Selected Papers on Renaissance Philosophy and on Thomas Hobbes*, ed. Piet Steenbakkers and Cees Leijenhorst. Dordrecht: Kluwer, 171–89.

—.2004d "Methodenfragen bei Spinoza und Hobbes: zum Problem des Einflusses," in *Selected Papers on Renaissance Philosophy and on Thomas Hobbes*, ed. Piet Steenbakkers and Cees Leijenhorst. Dordrecht: Kluwer, 45–71.

Schummer, Joachim. 2001. "Aristotle on Technology and Nature," *Philosophia naturalis* 38:1, 105–20.

Sepper, Dennis L. 1988. "Imagination, Phantasms, and the Making of Hobbesian and Cartesian Science," *Monist* 71, 526–42.

Shapin, Steven and Simon Schaffer. 1985. *Leviathan and the Air Pump: Hobbes, Boyle, and the Experimental Life*. Princeton, NJ: Princeton UP.

Shulman, George. 1989. "Metaphor and Modernization in the Political Thought of Thomas Hobbes," *Political Theory* 17, 392–416.

Skinner, Quentin. 1966. "Thomas Hobbes and His Disciples in France and England," *Comparative Studies in Society and History* 8, 153–64.

—.1972. "Conquest and Consent: Thomas Hobbes and the Engagement Controversy," in *The Interregnum: The Quest for Settlement 1646–1660*, ed. G. E. Aylmer. London: Archon Books, 79–98.

—.1978. *The Foundations of Modern Political Thought II: The Age of Reformation*. Cambridge: CUP.

—.1988a. "Meaning and Understanding in the History of Ideas," in *Meaning and Context: Quentin Skinner and His Critics*, ed. James Tully, rpt. Princeton, NJ: Princeton UP, 29–67.

—.1988b. "Political Philosophy," in *The Cambridge History of Renaissance Philosophy*, ed. Charles B. Schmitt and Quentin Skinner. Cambridge: CUP, 389–452.

—.1988c. "A Reply to my Critics," in *Meaning and Context: Quentin Skinner and His Critics*, ed. James Tully. Princeton, NJ: Princeton UP, 231–88.

—.1990. "Thomas Hobbes on the Proper Signification of Liberty," *Transactions of the Royal Historical Society* 40, 121–51.

—.1996. *Reason and Rhetoric in the Philosophy of Hobbes*. Cambridge: CUP.

—.1999. "Hobbes and the Purely Artificial Person of the State," *Journal of Political Philosophy* 7, 1–29.

Slomp, Gabriella. 2007. "Hobbes on Glory and Civil Strife," in *The Cambridge Companion to Hobbes's Leviathan*, ed. Patricia Springborg. Cambridge: CUP, 181–98.

Sorell, Tom. 2007. "Hobbes's Moral Philosophy," in *The Cambridge Companion to Hobbes's Leviathan*, ed. Patricia Springborg. Cambridge: CUP, 128–53.

Spragens, Thomas A., Jr. 1973. *The Politics of Motion: The World of Thomas Hobbes.* Lexington, KY: U. Kentucky Press.

Stillman, Robert E. 1995. "Hobbes's *Leviathan*: Monsters, Metaphors, and Magic," *English Literary History* 62, 791–819.

Strauss, Leo. 1963. *The Political Philosophy of Hobbes: Its Basis and Genesis,* trans. Elsa M. Sinclair. Chicago, IL: U. Chicago Press.

Sullivan, Vickie B. 1992. "'Machiavelli's Momentary 'Machiavellian Moment': A Reconsideration of Pocock's Treatment of the *Discourses*," *Political Theory* 20, 309–18.

Sunstein, Cass R. 1996. "Social Norms and Social Roles," *Columbia Law Review* 96, 903–68.

Tarlton, Charles D. 1996. "'The Word for the Deed': Hobbes's Two Versions of *Leviathan*," *New Literary History* 27, 785–802.

—.2001. "The Despotical Doctrine of Hobbes, Part I: The Liberalization of *Leviathan*," *History of Political Thought* 22, 587–618.

—.2002a. "The Despotical Doctrine of Hobbes, Part II: Aspects of the Textual Substructure of Tyranny in *Leviathan*," *History of Political Thought* 23, 61–89.

—.2002b. "Political Desire and the Idea of Murder in Machiavelli's *The Prince*," *Philosophy* 77, 39–66.

Townshend, Jules. 1999. "Hobbes as Possessive Individualist: interrogating the C. B. Macpherson thesis," *Hobbes Studies* 12, 52–72.

Tralau, Johan. 2007. "Leviathan, the Beast of Myth: Medusa, Dionysos, and the Riddle of Hobbes's Sovereign Monster," in *The Cambridge Companion to Hobbes's Leviathan*, ed. Patricia Springborg. Cambridge: CUP, 61–81.

Trentman, John. 1970. "Ockham on Mental," *Mind* 79, 586–90.

Tricaud, François. 1969. "'*Homo homini Deus*', '*Homo homini lupus*': *Recherche des Sources des deux Formules de Hobbes*," in *Hobbes-Forschungen*, ed. Reinhart Koselleck and Roman Schnur. Berlin: Duncker & Humboldt, 61–70.

—.1982. "An Investigation Concerning the Usage of the Words 'Person' and 'Persona' in the Political Treatises of Hobbes," in *Thomas Hobbes: His View of Man*, ed. J. G. van der Bend. Amsterdam: Editions Rodopi, 89–98.

—.1988. "Hobbes's Conception of the State of Nature from 1640 to 1651: Evolution and Ambiguities," in *Perspectives on Thomas Hobbes*, ed. G. A. J. Rogers and Alan Ryan. Oxford: Clarendon Press, 107–23.

Tuck, Richard. 1988. "Hobbes and Descartes," in *Perspectives on Thomas Hobbes*, ed. G. A. J. Rogers and Alan Ryan. Oxford: Clarendon Press, 11–42.

—.1993. *Philosophy and Government 1572–1651.* Cambridge: CUP.

Tully, James. 1988. "The Pen is a Mighty Sword: Quentin Skinner's Analysis of Politics," in *Meaning and Context: Quentin Skinner and His Critics*, ed. James Tully. Princeton, NJ: Princeton UP, 7–25.

Vanden Houten, Art. 2002. "Prudence in Hobbes's Political Philosophy," *History of Political Thought* 23, 266–87.

Vanderschraaf, Peter. 2001. "Instituting the Hobbesian Commonwealth," *Pacific Philosophical Quarterly* 82, 383–405.

Vatter, Miguel E. 2000. *Between Form and Event: Machiavelli's Theory of Political Freedom.* Dordrecht: Kluwer.

Vinciguerra, Lucien. 1999. *Langage, Visibilité, Différence.* Paris: J. Vrin.

Waldron, Jeremy. 2002. *God, Locke, and Equality: Christian Foundations of Locke's Political Thought.* Cambridge: CUP.

Wallace, William. 1984. *Galileo and his Sources: The Heritage of the Collegio Romano in Galileo's Science.* Princeton, NJ: Princeton UP.

Watkins, J. W. N. 1965. *Hobbes's System of Ideas: A Study in the Political Significance of Philosophical Theories.* London: Hutchinson University Library.

Whelan, Frederick G. 1981. "Language and its Abuses in Hobbes's Political Philosophy," *American Political Science Review* 75, 58–75.

Widner, Jennifer A. Summer, 1995. "States and Statelessness in Late Twentieth-Century Africa," *Daedalus* 124:3, 129–54.

Wilson-Quayle, James. 1996. "Resolving Hobbes's Metaphorical Contradiction: The Role of the Image in the Language of Politics," *Philosophy and Rhetoric* 29, 15–32.

Zarka, Yves Charles. 1984. "Espace et Représentation dans le *De Corpore* de Hobbes," *Recherches sur le XVIIe Siècle* 7, 159–80.

—.1987. *La décision métaphysique de Hobbes: Conditions de la politique.* Paris: J. Vrin.

—.1995. *Hobbes et la pensée politique moderne.* Paris: PUF.

—.1999a. "The Foundations of Natural Law," *British Journal for the History of Philosophy* 7, 15–32.

—.1999b. "The Invention of the Subject of the Law," *British Journal for the History of Philosophy* 7, 245–62.

Index